HUMAN AND ENVIRONMENTAL JUSTICE IN GUATEMALA

In 1996, the Guatemalan civil war ended with the signing of the Peace Accords, facilitated by the United Nations and promoted as a beacon of hope for a country with a history of conflict. Over twenty years later, the new era of political protest in Guatemala is highly complex and contradictory: the persistence of colonialism, fraught Indigenous-settler relations, political exclusion, corruption, criminal impunity, gendered violence, judicial procedures conducted under threat, entrenched inequality, as well as economic fragility.

Human and Environmental Justice in Guatemala examines the complexities of the quest for justice in Guatemala, and the realities of both new forms of resistance and long-standing obstacles to the rule of law in the human and environmental realms. Written by prominent scholars and activists, this book explores high-profile trials, the activities of foreign mining companies, attempts to prosecute war crimes, and cultural responses to injustice in literature, feminist performance art, and the media. The challenges to human and environmental capacities for justice are constrained, or facilitated, by factors that shape culture, politics, society, and the economy. Contributors to this volume include human rights activist Helen Mack Chang, environmental journalist Magalí Rey Rosa, former Guatemalan Attorney General Claudia Paz y Paz, as well as widely published Guatemala scholars.

STEPHEN HENIGHAN is a professor and head of Spanish and Hispanic Studies at the University of Guelph.

CANDACE JOHNSON is a professor in the Department of Political Science at the University of Guelph.

Human and Environmental Justice in Guatemala

EDITED BY STEPHEN HENIGHAN
AND CANDACE JOHNSON

UNIVERSITY OF TORONTO PRESS
Toronto Buffalo London

© University of Toronto Press 2018
Toronto Buffalo London
utorontopress.com

ISBN 978-1-4875-0389-5 (cloth)   ISBN 978-1-4875-2297-1 (paper)

Library and Archives Canada Cataloguing in Publication

Human and environmental justice in Guatemala / edited by Stephen
Henighan and Candace Johnson.

Includes bibliographical references and index.
ISBN 978-1-4875-0389-5 (cloth). – ISBN 978-1-4875-2297-1 (paper)

1. Social justice – Guatemala. 2. Environmental justice – Guatemala.

I. Henighan, Stephen, 1960–, editor II. Johnson, Candace, 1971–, editor

HM671.H86 2018   303.3'72097281   C2018-902366-X

This book has been published with the help of a grant from the Federation
for the Humanities and Social Sciences, through the Awards to Scholarly
Publications Program, using funds provided by the Social Sciences and
Humanities Research Council of Canada.

University of Toronto Press acknowledges the financial assistance to its
publishing program of the Canada Council for the Arts and the Ontario Arts
Council, an agency of the Government of Ontario.

Canada Council   Conseil des Arts
for the Arts   du Canada

ONTARIO ARTS COUNCIL
CONSEIL DES ARTS DE L'ONTARIO
an Ontario government agency
un organisme du gouvernement de l'Ontario

Funded by the   Financé par le
Government gouvernement
of Canada   du Canada

Contents

Acknowledgments

This book is a product of the extensive web of contacts in Guatemala and Canada forged by the University of Guelph / University of Saskatchewan Latin American Semester in Antigua Guatemala. Between 1997 and 2015, this academic semester abroad exposed Canadian students and faculty to Guatemalan scholars, activists, and artists. Three contributors to this volume served as coordinators of this semester, and others studied in or visited it. The connections between Canadian and Guatemalan researchers exemplified by this volume represent only part of the extensive series of cross-cultural dialogues created by this semester. The authors express their gratitude to the semester's founders, Kris Inwood of the University of Guelph and Jim Handy of the University of Saskatchewan.

In addition to this semester, the University of Guelph has hosted a number of events related to ongoing research efforts in and about Guatemala. In 2014, Helen Mack Chang was awarded an honorary doctorate and delivered the annual Hopper Lecture in International Development. Later that same year, we organized a workshop on Human and Environmental Justice in Guatemala, which served as the impetus for this book. Participants in this workshop included Claudia Paz y Paz, Magalí Rey Rosa, and Kirsten Weld. And in 2017 Victoria Sanford and Ivan Velásquez, commissioner of the International Commission Against Impunity in Guatemala (CICIG), visited the university to talk about impunity and justice. We would like to thank everyone who participated and provided their insights on these topics, including the speakers, and our students and colleagues.

Beyond campus, we would like to acknowledge the communities and individuals in Guatemala who are affected by the topics under consideration in this book. These include many of the activists and practitioners already mentioned, but extend to the countless other individuals who struggle for justice daily, such as Angélica Choc, Rosa Elbira Coc Ich, Germán Chub, Nery Rodenas, and Sara Curruchich Cúmez, as well as those who support them. Many of them have shared their time and ideas with us, and we are enormously grateful for that. For additional help in Guatemala, we're grateful to Vincent Simon and Julio Serrano Echeverría.

For funding assistance and support for on-campus events in Guelph related to this project, we thank the Community Engaged Scholarship Institute of the University of Guelph, the Political Science Department, International Development Studies, and the School of Languages and Literatures. We're grateful to Linda Hawkins, Elizabeth Jackson, Ana Lorena Leija, Ana María Méndez Dardón, and Shirley Shanahan for helping to organize such events, and to Kurt Annen, Lisa Maldonado, the late Kerry Preibisch, and Grahame Russell for enriching our conversations about Guatemala over the years. We thank Daniel Quinlan and the staff of the University of Toronto Press for their encouragement and support, and the Social Sciences and Humanities Research Council of Canada for the award of a grant through its Awards to Scholarly Publications Program.

Abbreviations

AGDH	Acuerdo Global sobre Derechos Humanos (Comprehensive Agreement on Human Rights)
ACI	Alianza Contra la Impunidad (Alliance against Impunity)
ALBA	Abraham Lincoln Archives
AIDA	Asociación Interamericana para la Defensa del Ambiente (Inter-American Association for Environmental Defence)
ASIES	Asociación de Investigación y Estudios Sociales (Association for Research and Social Studies)
AVEMILGUA	Asociación de Veteranos Militares de Guatemala (Association of Military Veterans of Guatemala)
CACIF	Comité Coordinador de Asociaciones Agrícolas, Comerciales, Industriales y Financieras (Coordinating Committee of Agricultural, Commercial, Industrial, and Financial Associations)
CALAS	Centro de Acción Legal Ambiental y Social de Guatemala (Guatemalan Environmental and Social Legal Action Centre)
CC	Corte de Constitucionalidad (Constitutional Court)
CEH	Comisión para el Esclarecimiento Histórico (Comission for Historical Clarification)
CGN	Compañía Guatemalteca de Níquel (Guatemalan Nickel Company)
CIA	Central Intelligence Agency (United States)
CIACS	Cuerpos Ilegales y Aparatos Clandestinos de Seguridad (Illegal Groups and Clandestine Security Organizations)

CICIG	Comisión Internacional Contra la Impunidad en Guatemala (International Commission against Impunity in Guatemala)
CIEL	Centre for International Environmental Law
CPR	Comunidades de Población en Resistencia (Communities of the Population in Resistance)
CRP	Comisión de Reforma Policía (Police Reform Commission)
CREOMPAZ	Comando Regional de Entrenamiento de Operaciones de Paz (Regional Centre for Training in Peacekeeping)
CSJ	Corte de Suprema Justicia (Supreme Court)
CSR	Corporate Social Responsibility
CUC	Comité de Unidad Campesina Peasant Unity Committee (Peasant Unity Committee)
DMI	Defensoría de la Mujer Indígena (Department of Protection of Indigenous Women's Rights)
EIA	Environmental Impact Assessment
EMP	Estado Mayor Presidencial (Presidential Guard)
ESI	Environmental Sustainability Index
FAFG	Fundación de Antropología Forense de Guatemala (Guatemalan Forensic Anthropology Foundation)
FAMDEGUA	Asociación Familiares de Detenidos-Desaparecidos de Guatemala (Association of Relatives of the Detained and Disappeared in Guatemala)
FAO	Food and Agriculture Organization (United Nations)
FAR	Fuerzas Armadas Rebeldes (Rebel Armed Forces)
FRG	Frente Republicano Guatemalteco (Guatemalan Republican Party)
FUNDESA	Fundación para el Desarrollo de Guatemala (Foundation for the Development of Guatemala)
GAM	Grupo de Apoyo Mutuo (Mutual Support Group)
IACHR	Inter-American Commission on Human Rights
IDPP	Instituto de la Defensa Publica Penal (Institute for Public Criminal Defence)
ILO	International Labour Organization
INACIF	Instituto Nacional de Ciencias Forenses (National Institute for Forensic Science)
INCO	Canadian International Nickel Company
KCA	Kappes, Cassiday and Associates
LCC	Ley de Compras y Contrataciones (Procurement and Contracting Law)

LCP	Ley de Comisiones de Postulación (Law for Nominating Commissions)
LEPP	Ley Electoral y de Partidos Políticos (Electoral and Political Party Law)
MEM	Ministerio de Minas e Energía (Ministry of Energy and Mines)
MINUGUA	Misión de Verificación de las Nacionas Unidas en Guatemala (United Nations Verification Mission in Guatemala)
MP	Ministerio Público (Public Prosecutor's Office)
MPJ	Movimiento Pro Justicia (Pro-Justice Movement)
OAS	Organization of American States
ODHAG	Oficina de Derechos Humanos del Arzobispado de Guatemala (Human Rights Office of the Archbishop of Guatemala City)
OIM	Organización Internacional de Mnigraciones (International Organization of Migration)
ORPA	Organización Revolucionaria del Pueblo en Armas (Revolutionary Organization of the People in Arms)
PAC	Patrullas de Autodefensa Civil (Civil Defence Patrols)
PBI	Peace Brigades International
PDAC	Prospectors and Developers Association of Canada
PGT	Partido Guatemalteco de Trabajo (Guatemalan Workers' Party)
PNC	Policía Nacional Civil (National Civil Police)
PP	Partido Patriota (Patriot Party)
REMHI	Guatemala: Nunca Más, Recuperación de la Memoria Histórica (Guatemala: Never Again, Recovery of Historical Memory)
SAT	Superintendencia de Administración Tributaria (Tax Administration Supervisory Office)
TSE	Tribunal Supremo Electoral (Supreme Electoral Court)
UDEFEGUA	Unidad de Protección a Defensoras y Defensores de Derechos Humanos de Guatemala (Guatemalan Human Rights Defenders Protection Unit)
URNG	Unidad Revolucionaria Nacional Guatemalteca (National Guatemalan Revolutionary Unity)
USAC	Universidad de San Carlos de Guatemala (University of San Carlos, Guatemala City)
WOLA	Washington Office on Latin America

PART ONE

Imagining Justice

1 Introduction: Transitional, Transnational, and Distributive Justice in Post-War Guatemala

CANDACE JOHNSON

In April 2015, the United Nations Comisión Internacional Contra la Impunidad en Guatemala (CICIG; International Commission against Impunity in Guatemala) revealed a major corruption scandal, one that was novel only in degree. The substance of the scandal, in this case a fraud scheme that enabled import businesses to avoid customs duties, is dishearteningly familiar in a country whose rates of taxation and social spending are appallingly low, even by regional comparisons (World Bank 2016a, 2016b). The scandal, known as "la línea," referred to a phone line that ran directly to the core of executive power, the offices of Vice-President Roxana Baldetti and President Otto Pérez Molina and their accomplices, who accepted millions of dollars in bribes in facilitation of a system of tax evasion that robbed the country of tens, maybe hundreds, of millions of dollars in revenue. The scandal broke in the context of an acute state of institutional failure and corruption, one that bordered on absurdity. The year before, in May 2014, Attorney General Claudia Paz y Paz (a contributor to this volume) was forced to resign in the wake of her now infamous prosecution of former president Efraín Ríos Montt for war crimes and genocide. The country's civil war (1960–96) has been followed by a fragile peace. The attempt to hold responsible the intellectual authors of the atrocities of the war constituted a critical step towards reconciliation. But the trial became a threat to established interests, and Paz y Paz and her ambitious prosecutorial agenda were ushered out of office. The removal of Paz y Paz echoed the Supreme Court's decision in May 2013 to repeal Ríos Montt's guilty conviction and roll back the proceedings to remedy the technical problem that was the ostensible reason for overturning the historic verdict. The trial, which resumed in 2014, was transformed into a judicial circus, with Ríos Montt, by then eighty-six years old and in poor health, wheeled into the courtroom for required appearances on a gurney. The criminal proceedings were suspended a number of times over the next two years, and reinitiated most recently in November 2016, when a Guatemalan judge ruled that Ríos Montt would face genocide charges again. This time, however, the former dictator would be tried by proxy through his lawyers, as he has been determined mentally unfit to stand trial (Democracy Now 2016).

While the 2013 verdict was a significant step towards justice, its immediate reversal is evidence that the powerful interests of the oligarchy, entrenched in the army and state, continue to prevail. But such an acknowledgment does not engender complacency. In reaction to "la línea," tens of thousands of Guatemalans gathered in the streets and congregated in Constitution Square to protest the latest and perhaps most unabashed political scandal of recent years. This "revolution of dignity" was fearless and relentless and eventually resulted in the resignation of President Pérez Molina. It also forced the withdrawal, in the midst of the 2015 electoral campaign, of presidential candidate Manuel Baldizón, aligned with the existing political and economic power structure, who had hitherto been considered a veritable shoo-in. The protestors favoured a "none of the above" presidential ticket and had proposed a citizens' congress in place of the usual executive authority. In the end, however, this option did not succeed, and the national elections resulted in a mandate for Jimmy Morales, an unlikely presidential candidate.

At first glance Jimmy Morales seemed to be the least threatening presidential option. He was well known as a television host and comedian with no obvious political ambitions. However, his links to the army became the focus of domestic aspersions, and his comedic insensitivities offended international sensibilities. In Guatemala he was chosen as the most anti-establishment candidate, an outsider who is supported by former military officials but whose own record is clean. He was regarded by many as the best of the bad options, in an election with a voter turnout of less than 50 per cent. His outsider status is critically viewed beyond Guatemala, as his comedy is seen as more offensive than frivolous. The *New York Times* reported that Morales's television shows regularly featured characters that included "one in blackface and a Japanese prisoner of war" (Malkin and Wirtz 2015), which suggests that there is little hope of moving forward on reparations in the context of racialized injustice.

Race is only one vector of injustice, yet it is of particular importance in the context of a country that by most estimates is more than 50 per cent Indigenous. It is this fact, in addition to the concentration of the Indigenous population in rural areas, that was the basis for the UN's determination that there had been genocide during the civil war, as found by its Comisión para el Esclarecimiento Histórico (CEH; Comission for Historical Clarification). This historical fact is a matter of deep controversy in contemporary Guatemala, as some believe that there was a genocide (*sí, hubo genocidio*), and that acknowledging this difficult fact is the first step towards justice; others believe that there was no genocide (*no hubo genocidio*) and that the genocidal label misunderstands the nature of the conflict and threatens to further divide a fractured society. President Morales is a staunch advocate of the latter camp (Elías 2016). The president's denial of the genocide has negative symbolic effects for those who were victims of state-sponsored violence and tangible consequences, as *no hubo genocidio* might mean prosecutorial reprieve for military officers.

Helen Mack Chang, Guatemalan human rights advocate and recipient of the Right Livelihood Award, explains in chapter 5 the contours of the struggle for justice and the importance of holding responsible the intellectual authors of state-sponsored violence. Similarly, environmental rights activist Magalí Rey Rosa (chapter 4), long-time *Prensa Libre* columnist and founder of Savia: School of Ecological Thought, documents the consequences of environmental injustices, related to international extractive industries, and the ways in which environmental harms are enacted through institutional and human injustices. It is no coincidence that the military targets of the civil war are also the areas that became sites for hydroelectric projects and gold and nickel mines. The scorched earth policies of Presidents Romeo Lucas García (1978–82) and Ríos Montt (1982–3), and carried out, in part, by President Otto Pérez Molina (2011–15), who, in his capacity as army general, led operations in the Ixil Triangle in 1982–3, served the economic interests of the established political class. And so, it seems, there has been and remains a morbid logic to the institutionalized violence and impunity, and fierce contestation over the facts and conclusions of Guatemala's recent violent past.

Prospects for Justice

Given the above details and analysis, it might seem that the quest for justice in any form is dishearteningly futile in Guatemala. As the contributors to this volume demonstrate, Guatemalan political, economic, and social life defies the most basic notions of fairness and human dignity. But there are multiple struggles towards this end and enough victories to inspire perseverance in the face of apparent futility. Much of the struggle for fairness and dignity is captured by the conceptual promise of transitional justice, which is a set of practices aimed at post-war healing and reconciliation. Common practices of transitional justice include truth commissions, historical memory recuperation projects in symbolic and real terms, which include the careful reconstruction of the national police archive and continued excavation of mass graves (Weld 2014), violent socio-political resistance, and relentless social criticism through art and other cultural expressions, such as that of Regina José Galindo and Rosa Chávez, the focus of chapter 7 by Rita Palacios. Guatemalans have engaged, and continue to engage, in all of these practices, often at great risk. Justice in transitional terms seems to be mostly aspirational, although each piece of the struggle serves to contribute *un granito* (to borrow a term from filmmaker Pamela Yates), one grain of sand that along with all of the others can cause the ground to shift in significant ways. In this way justice "victories" in the form of court cases, prosecutions, or international declarations of genocide, are signposts, and incremental, not "once and for all."

   This edited collection reinforces the argument presented by Elena Martínez Barahona, Martha Liliana Gutiérrez Salazar, and Liliana Rincón Fonseca that transitional justice is a process, each stage incomplete before the next begins. Often initial phases focus on recovery of historical memory (What happened? To whom? By whom? For what reasons?) and truth telling, which unravels slowly and in ways that reveal gender and racial oppressions (Martínez Barahona, Gutiérrez Salazar, and Rincón Fonseca 2012, 103; see also Crosby and Lykes 2011, 458). As recovery of historical memory reveals truths, the institutions of justice are put to the test through judicial proceedings (indictments, prosecutions). It is the latter (the procedural mechanics of justice) that reveal injustices that run deeper than the temporally confined episodes of genocide and civil war. Institutions are unable to handle the very crimes that they purport to address because their stewards are often the perpetrators. More directly, police chiefs, judges, army generals, and elected officials at all levels including the presidency have been both the material and intellectual authors of state-sponsored violence. Political and juridical institutions were weak prior to the war, enabling these protectors of the public trust to commit routinized violence with impunity and with the resources of the Guatemalan state, and the support of foreign (primarily American) aid, and so on. The patterns of impunity were established long before the civil war began in 1960 and can be traced back to the Conquest. Therefore, the real story is not one of Ríos Montt versus Attorney General Claudia Paz y Paz or Judge Yassmin Barrios, or the Ladino minority (non-Indigenous) versus the Indigenous majority, but Indigenous (pre-Columbian) cultures and livelihoods versus foreign exploitative interests and the globalized economy.

Martínez Barahona, Gutiérrez Salazar, and Rincón Fonseca describe two models of transitional justice that have been employed in Latin America. The first privileges the truth and exchanges retribution for recovery of historical memory (2012). In other words, truth commissions have been used as a way to find out "what happened" and offer perpetrators amnesty for their testimony. The second model does not make this exchange; rather, it uses the search for truth, also of utmost importance, as a means to test and thereby strengthen the institutions of justice. The second model is in practice in Guatemala. As several of the chapters reveal, there is a long and circuitous path to imperfect judicial outcomes. Some measure of justice was finally achieved in the case of Myrna Mack, who was assassinated 1990 in by officers of the Estado Mayor Presidencial (EMP; Presidential Guard). In 2003 the Inter-American Court of Human Rights identified Myrna's murder as a crime of the state (the direct perpetrator had been convicted of the assassination by a Guatemalan court in 1993); the state accepted its responsibility in the commission of the crime. Then in 2014 three of the ex-police officers implicated in assassinating the detective who investigated the Mack case were arrested. The first trial began in 2016, twenty-four years after the murder of the detective was committed. However, the "intellectual authors" of the crimes, the president or other high-ranking officials, have never been named. Similarly, in the Ríos Montt case, treated in more detail by Claudia Paz y Paz (chapter 6), the attorney general who prosecuted the former president and ex-military general, the trial and initial findings (of 2013) were monumental. This case was the first time a president had ever been convicted of genocide by the supreme court of his own country, as opposed to an international criminal court. But the verdict was immediately overturned and was succeeded by a raft of delays and stays of prosecution. In a similar vein, Magalí Rey Rosa (chapter 4) and Kalowatie Deonandan and Rebecca Tatham (chapter 3) explain the contradictions of the court cases that were the products of mining-related harms. The violence that is endemic to mining communities has found redress in both Guatemalan and Canadian courts, the trials unfolding with both inspiring and absurd qualities. Justice, as sought in Guatemala, might, in its simplest form, be tantamount to respect for basic human rights. There are, of course, more complex philosophical and political dimensions, which are the focus of this book. Notwithstanding the definitional machinations, to which I now turn, it is worthwhile to keep in mind, as George Lovell so eloquently demonstrates in chapter 9, that justice in any form is a daily struggle.

Theorizing Justice

As explained above, Guatemala has chosen the ambitious, more complex model to confront internal struggle. The quest for transitional justice, a conceptualization of justice that relates directly and specifically to the juridical and ethical dimensions of the movement of a society from a period of conflict to one of peace, is coterminous with the quest for institutional integrity. By putting fragile institutions to the test through inquiries, prosecutions, and political challenges, the state is forced to both confront its violent recent past and acknowledge the potential and limitations of its own institutions. There is much work to be done on a domestic scale in Guatemala to attempt social and institutional strengthening. The contributions in this volume address this requirement for justice. Claudia Paz y Paz explains in chapter 6 that one of the most important legacies of the Ríos Montt trial was "a strengthening of the rule of law." Such efforts to strengthen institutions and bring truth and healing to the victims respond to the troubling patterns of "internal colonialism" (Burgos-Debray [1985] 2010, xii), which are self-replicating and mutually reinforcing elements of oligarchy, corruption, violence, and impunity. At the same time, the contributors recognize the devastating qualities of external colonialism, which reveal the transnational character of political dysfunction and economic exploitation. Both logics, of internal and external colonialism, conspire to produce seemingly irreparable injustices.

With regard to the latter, externally located condition of justice, it is instructive to consider Iris Marion Young's "social connections model" of transnational justice and responsibility. According to Young, social connections (mutual responsibilities) are more important than the institutions of justice themselves. This way of understanding justice goes against much of the prevailing academic wisdom among moral and political philosophers as it assigns responsibility across borders. John Rawls and those who focus their analyses on the supremacy of territorially defined citizenship rights believe that citizens, members of a single political community or nation-state, have stronger obligations to one another than they do to other people in different nation states (Rawls 1999, 3–26; see also Miller 1995, 49). In *Responsibility for Justice*, Young explains that "ontologically and morally, though not necessarily temporally, social connection is prior to political institutions. This is the great insight of social contract theory. The social connections of civil society may exist without political institutions to govern them" (2011, 139). This also serves as a potential justification for the authority of non-state alternative "institutions," such as Mayan justice. Young elaborates her argument for a global conception of responsibility for justice with reference to Charles R. Beitz and Thomas Pogge. In short, Beitz and Pogge explain that the clear evidence of the existence of global institutions that secure rights and relationships, and facilitate transnational economic exchanges, defies the possibility of a model of justice that pertains exclusively or primarily to the nation state. Young aptly demonstrates that because there are global structures that both facilitate and constrain justice, it is impossible to reduce conceptual and practical pursuit of justice to the institutions and social relationships of the nation state (for a discussion of transnational civil society in Guatemala see Johnson 2016, 143–59). Her conclusion is that, "to the extent that agents, whether individuals or organizations, are indifferent to or support the operation of this global basic structure, they participate in the continuing enactment of injustice" (2011, 142). Put another way, Young argues that we all share global structures – economic and political, primarily – and that these structures produce predictable, interrelated injustices. Therefore, responsibility is shared across borders. This seems to be particularly appropriate for transnational relationships that have tangible and highly visible effects, such as those established by Canadian mining companies that operate in Guatemala.

Within the context of structural and transnational dimensions of justice and injustice, it makes sense to conclude that transitional justice and injustice also has a transnational character. Guatemala's protracted civil war was instigated and sustained in large part by foreign influences, and transitional remedies have been sought in many extraterritorial venues, often as a way to "leverage domestic processes" (Roht-Arriaza 2001, 83). Elin Skaar (2011, 2–4) and Cath Collins (2010, 21–35) have suggested that it would make sense to expand the conceptual parameters of transitional justice to include a "post-transitional" phase, which would address harms such as continued state-sponsored violence and human and environmental damage caused by mining committed after the conflict had officially ended. The arguments advanced in this volume support this proposition. The contributions would also suggest the need for consideration of what we might refer to as "pre-transitional justice" as a means of focusing on those forces that enabled the conflict in the first place.

However, time horizons are only one element of the difficulty in the quest for justice. The concept of time harbours a complex dynamic. Anika Oettler explains that the construction of truth narratives in the "present past" in Guatemala involves the collective remembering of multiple subjects and individuals (2006, 3–4). The separation of past from present is not an easy task, as both past and present are defined by moments and memories of suffering, and those moments and memories have evolving meanings and political purposes. In addition, truth commissions, such as the UN's CEH and the Catholic Church's *Guatemala: Nunca Más, Recuperación de la Memoria Histórica* (REMHI; Guatemala: Never Again, Recovery of Historical Memory) complicate these narratives by attempting to reveal truths that might speak more to international norms of justice than to local cultural interpretations (Oettler 2006, 5–11; Arriaza and Roht-Arriaza 2008, 152–4; Isaacs 2010, 251–8; May 2013, 496–8). The balance between global and local justice imperatives is delicate. Often the latter relies on the former and might not exist without it. However, there are particular cultural Indigenous sensibilities that remain, even when benevolently overpowered by transnational institutions of justice. Stephen Henighan explains in chapter 8 that the literary style of Guatemalan writer Rodrigo Rey Rosa was transformed as a result of his awakening to environmental injustices. This awakening has taken a more polemical form (and content) but has also oriented itself towards Mayan conceptions of justice, which go beyond and question basic conservationist assumptions. Reconciling universalist and Mayan conceptions of justice is of critical importance in a country where the Indigenous peoples exist in the contradictory reality of a majority population with (at best) minority rights.

Lieselotte Viaene explores the incompatibility of Indigenous culture and human rights norms through an ethnographic study of Maya-Q'eqchi' reactions to the government's reparations program. One of the Q'eqchi' participants explained, "It is difficult now. It is difficult for us to come together as one, because it has all fallen apart, yes. You could say that it is like an ants' nest that was destroyed or a beehive: they all go their own way, you cannot bring them together anymore" (2010, 13).

The reparations payments in Alta Verapaz as part of the government's reparations program (the focus of Viaene's study) or in Baja Verapaz as compensation for Chixoy Dam damages (discussed below) cannot repair the torn social fabric, but they can compensate for lost property, which is compatible with Mayan (and likely most other) conceptions of justice. This example demonstrates the central argument of this volume: that there is a complicated relationship between transitional and distributive justice, yet the two conceptions are fused. Further, there are irresolvable tensions between the symbolic and the material elements of justice (Oettler 2006, 10), between truth and reparations. Naomi Roht-Arriaza explains that "a major critique of TJ [transitional justice] is that it has not engaged well with distributive justice, with socioeconomic rights. This is often cited as one reason why the results of TJ, and of transitions generally, have felt so limited from the perspective of people on the ground" (2001, 82).

The Justice Dilemma: Post-War Reconciliation and Distributional Imbalances

Guatemala is a country with high levels of both absolute and relative poverty; in other words, it is a poor country with high levels of inequality. The World Bank identifies Guatemala as a lower-middle-income country, yet also estimates that almost 60 per cent of the population lives in poverty (World Bank 2016b). In its country overview, the World Bank aptly describes this contradiction: "Thanks to prudent macroeconomic management, Guatemala has been one of the strongest economic performers in Latin America in recent years, with a GDP growth rate of 3.0 percent since 2012, and nearly 4.0 percent in 2015. Even so, it is one of the few countries in the region where poverty has increased in recent years, from 51 percent in 2006 to 59.3 percent in 2014" (World Bank 2016a). The economic growth, according to this overview, is largely attributable to the expansion of international trade and investment following the Peace Accords in 1996, and inequality is manifest in failing social indicators, such as high rates of maternal and infant mortality. This contradiction is the result of a combination of predictable Central American dynamics, replete with low rates of public spending, oligarchic tendencies, political instability, and fragile democratic institutions. Therefore, in discussions about justice in Guatemala, it would seem perfectly reasonable, for practical reasons and with the aim of improving the conditions of daily life, to limit the focus to distributional ethics and the possible ways of bringing about a fairer distribution of social, political, and economic resources.

This diagnosis defined the moment of democratic opening in Guatemala from 1944 to 1954, the "Ten-Year Spring" of the October Revolution. Democratic reforms began to emerge with the end of President Jorge Ubico's reign as a neocolonial, pro-American, military dictator. Ubico had held power since 1931 and resigned in 1944 in response to protest led by students from Universidad de San Carlos de Guatemala (USAC; San Carlos University) in Guatemala City. There had been growing unrest in response to gratuitous political corruption at all levels and to the state-sponsored violence that maintained it. The resignation of Ubico was followed by a brief transitional period of military rule, which led to the country's first democratic elections. These elections brought to power the relatively moderate Juan José Arévalo. In 1949 the Guatemalan military attempted to remove Arévalo but was unsuccessful. Arévalo fulfilled his mandate and was replaced by Jacobo Árbenz Guzmán in 1951.

The democratic opening that began in 1944 was fully realized through the efforts of Árbenz. As president he instituted a program of land reform and nationalized key segments of the economy. While this frustrated foreign (primarily American) economic interests in the country, the bigger problem from a U.S. perspective seemed to be the degree to which Árbenz was sympathetic to the communist-oriented Partido Guatemalteco del Trabajo (PGT; Guatemalan Workers' Party). Árbenz was not a member of the PGT, nor a declared communist, although he clearly had a social democratic agenda and had legalized the PGT, which had hitherto been outlawed. He also had several close confidants who were PGT members, including his wife, María Cristina Vilanova Castro de Árbenz. In the context of Cold War fears that had been heightened by recent revolutionary activity in Cuba in 1953, and ensuing panic about further communist threats in Latin America, Árbenz had surrounded himself with many whom Washington considered by to be subversives and seemed unwilling to assuage American political anxieties. CIA archives reveal that President Eisenhower had been planning for military action since 1952 by collecting names of suspected communists (Grandin, Levenson, and Oglesby 2011, 242–4) and destabilizing popular support for Árbenz (see, for example, Cullather 2011). The military descended upon the presidential palace in 1954, and a CIA-backed coup swiftly removed Árbenz and ended Guatemala's democratic moment (for a full examination of the coup, see Schlesinger and Kinzer 1982). What followed (1954–60) was little more than a succession of military governments that were committed to social control and oppression. The Cold War hysteria continued, and the "communist threat" became the motivation and justification for state-sponsored violence. In 1960 the violence transformed into a full-fledged civil war, with the Guatemalan Army, in addition to the police force, committing unimaginable atrocities against suspected communists and their mostly Indigenous sympathizers. The civil war officially ended in 1996 with the signing of the Peace Accords. In 1999 the final report by CEH, *Memoria de Silencio* (the truth commission was a requirement of the UN's negotiated peace), determined that more than 200,000 people had been killed during the war, and that the state had committed acts of genocide (United Nations 1999, 20). Much of what has happened in transitional justice (from 1996 to present) has centred on a country coming to terms, or not coming to terms, with its recent genocidal past.

As the chapters in this book demonstrate, Guatemala is a complex country, with complex historical narratives woven into complex political, social, and economic realities. The problems of transitional justice exacerbate distributional inequities, which result in deep inequality that is both worsened and facilitated by human rights violations. One example of this overlapping effect is seen in Río Negro, an extraordinarily beautiful and peaceful community that was devastated during the civil war. In 1978 the government of Guatemala authorized construction of the Chixoy hydroelectric dam, financed in large part by the World Bank and the Inter-American Development Bank, which today provides electricity to more than 2,500 rural communities in Guatemala (INDE 2015), although few households in the immediate Río Negro area have access to electricity. The plan to "develop" this particular area and create progress for the country in general involved relocating the local Indigenous residents, who lived in the fertile river basin, to new urban settlements. Some of the residents agreed and moved to Pacux, a "model community," which is basically a grid-like configuration of cinder block houses with no land for farming or river for fishing. The fate of these people was dislocation, poverty, and unemployment (see Rights Action 2005). It was dismal, but far better than the fate of those who refused to leave Río Negro.

The government, eager to move forward with dam construction, flooded the river basin and drowned those who had not already retreated into the mountains. Then it deployed the army to find those who remained. This happened at the height of the scorched earth campaign (1980–2), so the military was already mobilized for absurd levels of violence in search of "subversives." The forced evictions resulted in five massacres, in Río Negro, Xococ, Pak'oxom, Los Encuentros, and Agua Fría. More than 400 died (Dearden 2012). The survivors continue to tell the story and fight for justice (Chen Osorio 2009; Tecú Osorio 2006), which has also followed a frustrating and predictable path. In 2014 the U.S. Congress passed the Consolidated Appropriations Bill, which ordered reparations for the harms caused to the community (Russell 2014). In 2015, the government of Guatemala under President Pérez Molina made further rhetorical commitments to paying reparations and even went so far as to include the Chixoy dam reparations as a line item in the national budget (Barillas 2016). But the community has still not seen repayment and there is little reason to believe that the new government of Jimmy Morales will advance the funds.

This pattern of economic development project, resistance, and violence was repeated throughout the civil war period and continues today. El Estor, a community in the eastern part of Guatemala on the shores of Lake Izabal, was the site of military activity during the civil war and was suspected of being a base for guerrilla insurgency (Klippensteins 2016). With full recognition of this volatility, INCO, a Canadian mining company, began exploring the possibility of mining concessions and extraction in 1960, the year that the civil war officially began. In 1965 INCO received a "40-year mining lease to an area 385 square kilometres in size near El Estor from the Guatemalan government. In order to allow for open-pit mining, which was previously prohibited by the constitution, the military government suspends the constitution, dissolves congress and passes a new mining code" (ibid.). Ownership of the mine has changed several times, from INCO to Skye to Hudbay, all of which were Canadian companies. Hudbay was the foreign owner of the mine from 2008 to 2011 and operated through its local subsidiary, Compañia Guatemalteca de Niquel (CGN; Guatemalan Nickel Company). The Fénix nickel mine and processing facility are now owned by the Solway Group, a Russian multinational, but continue to be operated by the CGN, which maintains deposit rights. Yet despite changes in foreign ownership, the reality for local residents has remained the same. The stakes are high in mining communities, perhaps more so than for hydroelectric projects, as the former tend to be labour intensive, whereas the latter are not. This means that the lure of employment in places like El

Estor is a compounding complex factor in assessing mining-related opportunities and harms, as it asks community residents to choose between unending poverty and environmental devastation. The promise of a job creates support for, and resistance to, the mine, often among members of the same family. The lack of industrial and environmental regulations, which is endemic in Guatemala, means that communities are affected by soil and water contamination, round-the-clock heavy truck activity and blasting, livelihood destruction, and forced evictions.

Forcibly removing people from their land is by definition a violent act that contravenes the most basic human rights. But in Guatemala the violence escalates to extreme levels, because it is inflicted in a context of structural violence and impunity. This is demonstrated through the liability cases that have found their way to Canadian courts. In one community within the Fénix concession, Lote Ocho, eleven women were allegedly raped and their property destroyed and homes burned to the ground by "soldiers, police officers and mining security officials" (Daley 2016). The negligence suit of Margarita Caal Caal and ten others from Lote Ocho draws attention to the use of sexual violence to secure economic interests. It also reveals the difficulties of addressing reparations for sexual and reproductive violence. Colleen Duggan, Claudia Paz y Paz Bailey, and Julie Guillerot argue that while criminal proceedings to bring perpetrators of sexual and reproductive violence to justice are important, they neglect the narratives and collective needs of the victims. Thus the retributive and restorative elements of justice are also intertwined, creating more complexity and difficulty, as well as possibilities for redress. As the authors explain, "The question of how to repair the harm done cannot be delinked from the questions of why the harm was done and to whom" (2008, 193). In other words, justice must be sought in the "past present" and requires collective redress, not just prosecution of individual perpetrators.

In another local community, El Estor, the cases of Adolfo Ich and Germán Chub reinforce the political nature of the violence. Ich was a community leader and resister who was allegedly murdered by the head of security for the mine, Mynor Padilla, and Chub, permanently disabled, was a bystander. Chub's fate serves as a reminder that it is virtually impossible to remain unaffected by the violence that permeates mining communities. The Amended Statement of Claim of Caal vs Hudbay Minerals Inc. makes explicit connections between the immediate, mining-related violence and the political violence of the past civil war. The claim states,

> 34. There are currently several indigenous Mayan Q'eqchi' farming communities located on a small portion of the Fenix Property (the "Contested Land"). During the period relevant to this lawsuit, Skye Resources and CGN claimed that they had valid legal right to the Contested Land, while Mayan Q'eqchi' communities claimed and continue to claim that they are the rightful owners of the lands which they consider to be their ancestral homeland. The Mayan Q'eqchi' further claimed that any apparent rights to the Contested Land claimed by Skye or CGN are illegitimate as these rights were first granted by a dictatorial military government during the Guatemalan Civil War, at a time when Mayan Q'eqchi' communities were being massacred and driven off of their land.
>
> 35. According to the United Nations' sponsored truth commission, Comisión para el Esclarecimiento Histórico (the "Truth Commission") Mayan populations were particularly targeted during the Civil War, resulting in the "extermination, en masse, of defenceless Mayan communities purportedly linked to the guerrillas – including children, women and the elderly – through methods whose cruelty has outraged the moral conscience of the civilised world."
>
> 36. In 2006, an agency of the United Nations ruled that Guatemala had breached international law by granting mining rights to the Fenix Project to CGN and Skye without adequately consulting with local Mayan Q'eqchi' communities. The Guatemalan government and Skye Resources have ignored this ruling. (Ontario Superior Court of Justice 2012)

These passages also make abundantly clear the relationships between transitional, distributive, and transnational elements of justice. In order to establish the conditions for transitional justice, it is imperative to recognize the transnational character of the conditions that led to the harms. While it was the Guatemalan government that suspended the constitutional right to assembly in order to eliminate protest against the mine and ensure stability in the region (Imai, Maheandiran, and Crystal 2014, 287), and passed legislation that would authorize open-pit mining, it was INCO's engineers who rewrote the mining code (287). Moreover, this was not simply one international business transaction gone awry. Mining was a priority of the Canadian government, and therefore Canada was an active participant in the human and environmental devastation that ensued. "The history of INCO in Guatemala shows that Canadian mining interests were promoted by the Canadian government, and yet that the Canadian government did not take the initiative to address corporate accountability for the violence associated with these mining operations" (287). Canada has been a leader in human rights protection around the globe and seems unequivocal in its purposes in ensuring global justice. Unfortunately, it appears that this moral authority is impaired by layered economic interests. The promotion of extractive industry during the conflict seems dubious and highly opportunistic.

Regardless of its motives, it is important to understand that the Canadian government was and is not merely a passive regulator of corporate activities abroad; it has had and continues to have a political agenda that is characterized by both action and inaction. Responsibility for justice extends to Canada, especially in the mining sector. Canadian corporations control approximately 70 per cent of mining operations in Guatemala (Logan 2014), yet there appears to be no pressure by the Canadian government, regardless of the party in power, on the Guatemalan government to address its human rights record, enforce environmental regulations, or pay promised reparations. All of these improvements would facilitate greater distributive justice and reduce socio-economic inequalities. The Peace Accords seem to require greater distributional equity as a condition of transitional justice, in particular concerning education and recognition through education of the causes and consequences of the civil war. There are mechanisms in place to assign responsibility for justice (laws, courts, institutions, corporate social responsibility codes, international oversight), but while each independent mechanism might function effectively on its own or intermittently, there is no reliable system, national or transnational, that can be engaged to produce consistent and fair results.

In this context, what are the prospects for justice in Guatemala? On the face of it, there might be little reason for hope. But upon deeper investigation, there is hope in many places, some of them unlikely. One example is the groundbreaking precedent of standing for Guatemalan plaintiffs in bringing cases against Hudbay in the Canadian court system. While there is no real expectation that victories will be paid out in the lifetimes of the plaintiffs, the symbolic importance of the cases cannot be overstated. It signals that the lives of Indigenous Guatemalans are worthy of protection and that corporations will not be able to take impunity for granted any longer. In addition, it provides the victims with recourse and some vindication for their version of historical events.

The harms did happen and the foreign companies might be responsible. This is a powerful message to foreign mining companies, their governments, and the Guatemalan state. It lends legitimacy to narratives of victimization and impunity, which is fundamental to transitional justice pursuits. But this path to justice is also very costly. It further divides fractured communities. Many of the members of the community of Angélica Choc (Adolfo Ich's widow), La Unión, have come to regard her as an undeserving celebrity and are suspicious that she has profited financially from the court case. The tension surrounding her status has caused her to move out of the immediate area into a neighbouring town. While she has not received a payout and the idea of a settlement in her favour is mostly fantasy, she has been transformed into something that she was not before. She travels to Canada regularly and receives international delegations in El Estor. Her picture appeared in April 2016 in the *New York Times*. She is one of the bees that cannot go back to the hive.

In addition to the limited prospects for success of the cases, despite their symbolic merit, and the costs to individuals and communities, it is also important to recognize that these cases will do little, if anything, to address larger distributive issues. This would require greater acts of domestic and international political will than have typically been practised. In Guatemala, distributive justice would require greater political resources directed to social rights to public goods, like education and health care. The political assumption is that Guatemalan revenues are not sufficient to provide these services. But the evidence does not support this assumption. As noted, Guatemala's economy is strong relative to its social spending commitments. Moreover, the government, through the offices of the president and vice-president, seemed to have no qualms about squandering millions of dollars through "la línea" and numerous other scandals that destroyed health and the environment in order to deliver kickbacks to already unimaginably wealthy public officials. For example, in 2015 Vice-President Baldetti authorized 137.8 million quetzales (approximately US$17 million) to pay for chemical treatment for the polluted Lake Amatitlán, just outside of Guatemala City. But, upon testing, the decontaminant – "a magic formula" – was deemed ineffective, yet the revenue was never recovered (Rodríguez 2015; Balcárcel 2015; Reynolds 2015). Also in 2015, prosecutors laid charges against high-ranking officials, some of whom were relatives of a Supreme Court judge, in relation to a scandal at the Social Security Institute. Journalist Sophie Beaudoin reports, "According to prosecutors, the social security institute fraudulently awarded a US$15 million contract to the Pisa pharmaceutical company, at more than 15 percent commission, to ensure renal dialysis to social security patients though Pisa had neither the staff nor technical capacity to fulfil the contract. Prosecutors allege that at least ten patients died as a result of the illegal contracting, and many more suffered health complications" (Beaudoin 2015). These cases serve as further evidence that corruption and impunity are significant impediments to social rights, regardless of the facilitating or constraining effects of economic factors.

With regard to international corporate social responsibility (CSR), the trend appears to continue in the direction of voluntary corporate codes rather than public policy as such. Therefore, there is little incentive for Canadian companies operating abroad to make greater efforts to respect the livelihoods of affected communities, guarantee human rights, or ensure environmental integrity. In fact, it seems that the court cases and other forms of international activism solicit greater resistance to any genuine expression of CSR. In addition to the Hudbay cases in Guatemala, a precedent-setting case was brought against Chevron for harms, related to severe environmental pollution, committed in Ecuador. Imai, Maheandiran, and Crystal describe what is at stake in transborder pursuit of justice: "In a globalised world, encouraging ethical behaviour cannot be left to a single jurisdiction or a single institution. We hope that the time has come for Canadian courts to begin to participate in creating the mechanisms necessary to close the gap in corporate accountability." A spokesperson for Chevron, referring to the case in Ecuador, stated, "We're going to fight this until hell freezes over. And then we'll fight it out on the ice" (Imai, Maheandiran, and Crystal 2014, 299). In the context of these examples, justice might best be conceptualized as a process rather than an achievement. It is also the complexity of historical narratives, post-war reconciliation, transnational economies and legal systems, and repeated revelations of distributional inequities as a way of making sense of what has happened in the past and finding hope for what might happen in the future.

Human and Environmental Justice in Guatemala

This book focuses on "human" rather than "social" justice, because the former indicates the importance of human agency and action whereas the latter, more familiar term seems to absorb responsibility for justice into an imprecise collective subject. Human actors have ordered genocides and other atrocities, issued mining concessions, pursued corporate interests; they have also suffered human rights violations, created resistance, and offered solutions to endemic problems. The book's environmental focus captures both the social and physical elements of environmental justice. Taken together, human and environmental justice reveal what is at stake in complex transitional justice projects in post-war Guatemala. The chapters in this book investigate the possibilities for justice in Guatemala in the context of serious challenges to structural violence, weak institutions, culture of impunity, patterns of internal and external colonialism, and ethnic and gender injustice. It joins a robust literature that seeks to address the same issues but is unique in four important ways. First, it presents the link between human and environmental aspects of justice as primary. In order to understand the challenges of seeking justice in Guatemala, it is necessary to understand the relationship among individuals, communities, and natural resources. Many of the chapters demonstrate this acutely with investigations of violence and displacement (Nolin, Rey Rosa, Deonandan and Tatham, Henighan), and others illustrate the relationship more obliquely (Mack Chang, Paz y Paz, Palacios, Lovell), but all reveal the integration of human and environmental exploitation. Second, this book focuses on the complexity of the multiple processes of justice that exists in and for Guatemala. Transitional justice is central to other conceptions, as the reality of the civil war, from the long history that led to revolution through the genocide and its aftermath, affects virtually every aspect of political and social life in that country. Other orientations towards justice, such as distributive concerns that address inequality, are inseparable from transitional issues yet have their own internal dynamics. Guatemalan governments, such as that of Álvaro Colom (2008–12), which implemented a national social assistance program, have addressed distributive injustice independently of transitional efforts. However, the impossibility of sustained redistribution must be explained with reference to the historical complexity of transitional justice. Third, this book focuses on Canadian and Guatemalan perspectives on justice. In so doing it asks sharper questions about responsibility for justice in the context of particular political relationships. Canada has been heavily involved in the politics of the region through a variety of relationships among governments and non-governmental organizations and extractive industries, which have caused unimaginable damage. Likely Canada's role is of even greater significance in a post–Cold War context. With waning U.S. hemispheric hegemonic dominance from its Cold War peak, there is greater opportunity for Canada to play a more important role in Central America, which has both positive and negative effects. Canada is also involved in redressing harms of the past in the form of precedent-setting court cases and transnational activism. Fourth, and perhaps most importantly, this book makes the three contributions detailed above through the experiences and analyses of both internationally recognized human rights practitioners (Rey Rosa, Mack Chang, Paz y Paz) and leading academics (Nolin, Deonandan and Tatham, Palacios, Henighan, Lovell) from multiple disciplines and through diverse methodological approaches. The insights and testimony of the practitioners are critical to the analysis advanced by the academics. Together, they test the veracity of historical, political, and gendered narratives as a means to reveal truths and seek justice.

The Chapters

The book is divided into three sections: imagining justice, justice in practice, and cultural responses to injustice. The first section, which includes this introduction, is devoted to conceptual yet engaged discussions of justice by academic contributors. Chapter 2 is written by Catherine Nolin, a geographer at the University of Northern British Columbia who has been conducting research in and about Guatemala for more than two decades. In this chapter, Nolin examines the structural elements of violence and the pursuit of justice through a variety of ethnographic methods. She writes "from a place of witnessing," which provides both social scientific distancing and acknowledgment of embeddedness. Her work incorporates and reflects the work of prominent human rights practitioners and organizations such as the Fundación de Antropología Forense de Guatemala (FAFG; Guatemalan Forensic Anthropology Foundation) in Guatemala and beyond, and in so doing demonstrates that engaged scholarship has much to offer debates on transitional justice and global responsibility. Nolin's research on the importance of historical clarification, specifically the return of remains to the families of the dead and disappeared, speaks directly to the significance of post-transitional justice in Guatemala. Her powerful research reveals that "contemporary Guatemalan issues cannot begin to subside or to reverse until Guatemalans heal from the shattering of their communities, until justice is evident, until the bones of their loved ones are returned to them (Nolin Hanlon and Shankar 2000). Surviving family members want the return of more than 200,000 murdered and more than 45,000 'disappeared': those people strategically removed from society, or forcibly disappeared, for their ideals, intellect, political stance, and vision for a better Guatemala" (Nolin, this volume, 37). Nolin's research also emphasizes the transnational nature of justice and injustice. She explains that "through this research we attempt to understand a place like Guatemala in the context of North-South connections, through the scales of accountability (from local through to the North and South), and the necessity to bring an historical perspective which asks what is our responsibility in the North in creating and understanding Guatemala as an exploited and abused country rather than a 'developing country.'"

In chapter 3 Kalowatie Deonandan and Rebecca Tatham examine the combined effects of racialized gender injustice and mining harms. The theoretical analysis and empirical findings that they present reinforce the narratives in chapters by Magalí Rey Rosa, Helen Mack Chang, and Claudia Paz y Paz, and confirm David Carey's admonition that the Guatemalan judicial system "was both an extension of the government's coercive and administrative apparatus and, ironically, one the few escape valves available for popular dissent" (Carey Jr 2013, 15). The focus of this chapter is the criminal case against Mynor Padilla, former head of security for CGN, which operated as Hudbay's subsidiary in Guatemala, and its relationship to the civil suit against Hudbay in Canada. Through their research (extended case empirical inquiry in the eastern region of Guatemala), we hear from Angélica Choc, the widow of Adolfo Ich. This is powerful testimony, which reveals the complexity of justice that is the subject of this book. Choc explains that, "with respect to this criminal case, I am very appreciative of the international support, specifically international observers who have accompanied me to this trial. For me this is very important and I am very grateful of their presence because that way they realize what justice really looks like in Guatemala" (Deonandan and Tatham, this volume, 65). Her words emphasize the importance of international witnesses who provide solidarity support in her struggle and embody the chapter's main argument, that "various forms of solidarity, from the global to the local, have empowered Choc in her struggle for justice." The authors arrive at this conclusion by considering the transnational solidarity activities of three organizations: Rights Action and AcoGuate, both NGOs, and CICIG, an appendage of the UN. They further conclude that the various roles of these organizations are essential to Choc's case, and to the larger national project of transitional justice, yet seriously restricted in the context of global capitalist ventures and local impunity, discrimination, and violence.

The second section of the book is an examination of the pursuit of justice by key Guatemalan practitioners and activists. The extent of the mining exploitation and abuse that is introduced by Deonandan and Tatham is carefully and thoroughly documented in chapter 4, written by Magalí Rey Rosa, who is one of Latin America's most prominent environmental activists. Her chapter is a condensation of her articles in *Prensa Libre*, Guatemala's most widely read newspaper. The journalistic account provided by Rey Rosa begins to reveal the political complexity of seemingly simple problems. Beginning in 2004, she tracks the evolution of Canadian extractive industry in Guatemala and exposes the contradictions between political promises and economic imperatives. She explains that justice in Guatemala is fused with the problem of land distribution, and that human and environmental justice are inseparable. Rey Rosa's writings are dishearteningly prophetic and refreshingly frank. In the fall of 2005 she wrote, "If Montana Mining, as the subsidiary of Canada's Glamis Gold, represents the new generation of mining, supposedly modernized, with cutting-edge technology, which acts in a responsible way towards people and their land, then we're screwed" (Rey Rosa, this volume, 93). She explained that, by the end of that same year, "The government has announced that, due to official budget cutbacks, some communities in Sipacapa will have no schoolteachers in 2006. The solution suggested by government officials is that the communities accept schoolteachers chosen and paid for by Montana Mining. Could this be an official reprisal for the community's rejection of the mining project?" (94). This exposes the relationship among transitional, transnational, and distributive justice in the most horrifying and efficient way. She also examined the important recurring theme of gender justice. In March 2012 she wrote of women from mining-affected communities who had come to the capital to protest, "They march under the sun, day after day, in the hope that their gesture will move us" (106). The chapter is a compelling marshalling of evidence on human and environmental injustices, the need for mining companies and the governments of Guatemala and Canada to take responsibility, and the layers of impunity that render justice and responsibility extremely unlikely achievements in any form. It is also a unique opportunity for Canadian readers to hear the story of their mining operations from a Guatemalan perspective.

The Guatemalan institutions that facilitate exploitation are carefully described in chapter 5 by Helen Mack Chang, human rights activist, founder of the Myrna Mack Foundation, and recipient of the Right Livelihood Award (1992) for her tireless search for justice for her sister, anthropologist Myrna Mack Chang, who was assassinated by the EMP in 1990. She explains that the military apparatus that was so efficient and violent during the protracted civil war has "mutated to become organized crime networks, with participants of every sector of society, both public and private.... Their activities, which have now gone well beyond contraband and illegal trafficking, have consequences within political power structures" (Mack Chang, this volume, 119). These networks reach into industrial sectors (including mining, African palm oil production, hydroelectric projects) and into the Congress, the judiciary, and the presidency. The layers of corruption and a culture of impunity interact to produce environmental harms, violence, and structural injustice, which have recently reached crisis proportions. In spite of widespread unrest evidenced by the massive political protests of 2015, Mack Chang concedes that "we are living in a failed political regime, in a culture where impunity and corruption are slowly gaining active social consensus. This consensus makes it possible to legalize the illegal, and obtain social and political power for a state that has been completely co-opted in service of organized crime and has stepped away from seeking the public good, the fundamental goal of democracy" (121). In a powerful metaphor, Mack Chang explains that justice, once the "Cinderella" of the state because she works tirelessly with no recognition or effect, is now more akin to a prostitute who will serve those who pay her well.

Chapter 6 is a first-person narrative of the current state of the judicial and legal institutions that facilitate or constrain the advancement of human rights and justice in Guatemala. It is written by Claudia Paz y Paz, who served as attorney general from 2010 to 2014, during some of the country's most turbulent post-war times. Paz y Paz describes the creation of the CICIG and other judicial institutions, and the subsequent actions taken by victims of crime and impunity to stand up for their rights, which ultimately culminated in a criminal trial against the intellectual authors and perpetrators of genocide and crimes against humanity. Former president Efraín Ríos Montt was among those tried, and the final ruling convicted him of genocide. Although the ruling was quickly struck down, the trial and the events that preceded it represent a huge step forward in strengthening the rule of law. Her analysis emphasizes the importance of the practice-focused

model of transitional justice. Through identifying crimes and perpetrators, bringing cases to trial, securing legal venues, and prosecuting offenders, the institutions of justice are fortified. The processes are
always frustrating and the outcomes often abominations, but they demonstrate well the tenacity of the officials and citizens who value justice
enough to pursue it at all costs. They also harbour the potential for genuine transformation through truth and recovery of historical memory.
Paz y Paz describes the powerful testimony of the ten Indigenous
women who had survived sexual assault by the army: "There, in their
own language, they faced Ríos Montt, the man who had ordered that all
these human rights be committed; faced the man who commanded the
armed forces and who had once had all the powers of the state" (Paz
y Paz, this volume, 133). For Paz y Paz this is evidence of the amazing
power of justice to equalize, as "the trial finally made it possible, for
once, for the perpetrator and the victims to meet on equal footing and
for all to experience the equalization brought by justice."

Rita M. Palacios, professor of Spanish at Conestoga College, resumes
the discussion of historical memory and gender justice in chapter 7. This
chapter begins the final section of the book, which is a series of cultural
responses to injustice. While Palacios's research on the performance art
of Regina José Galindo and the poetry of Rosa Chávez effectively reinforces Paz y Paz's admonitions about justice, Palacios does not share
the former attorney general's faith in the practice of justice as a means
to strengthen the rule of law. In contradistinction, Palacios mocks "the
state's theatrical enactment of the rule of law, which is shown to rely
on artifice for its own legitimization, and violence for its enforcement"
(Palacios, this volume, 140). Thus the work of Galindo and Chávez "confronts the state's inability to bring about justice and initiates a process
of memory and reconciliation," and at the same time provides a means
for expression of injustice by performing and the suffering and violence
of society on and through the body. Palacios explains that "these formal
processes [of truth commissions and reports] categorize and archive the
individual and collective memories that can be *recorded* but overlook
what can only be *experienced*. In other words ... while reports, trials, and
sentencing may bring justice, they are not enough to address the lasting
effects of violence" (139). Performances of violence through provoca-
tive art provide a means to shatter the routinized responses to violence
and force audiences to deal with the discomfort of enduring trauma.
Palacios's research also reveals the racialized dimensions of violence
in Guatemala and the ways in which artists and activists can achieve

solidarity and justice through "Mayan" resistance. Further, in explaining the character of Chávez's poetry, she asserts that "the poetic voice sets in motion a decolonizing act that rejects the status quo and transmutes power and authority, showing humankind to be part of a much broader system that incorporates the natural world."

The relationships between environmental and human justice and between Mayan resistance and solidarity are central to the work of Rodrigo Rey Rosa, one of Guatemala's most internationally acclaimed authors. In chapter 8, Stephen Henighan, professor of Hispanic studies at the University of Guelph, examines the trajectory and impact of Rey Rosa's later, more political writings. He explains that Rey Rosa's environmentalism, which developed in the 1990s, inspired the author to reorient his literary focus. The novels *Lo que soñó Sebastián* (1994), *El material humano* (2009), and *Los sordos* (2012) are concerned with aspects of the natural world, Guatemalan history, and conceptions of Mayan justice. The kidnapping of Rey Rosa's mother in 1981 and his immersion in the literary avant-garde might have resulted in his physical and intellectual distancing from his native Guatemala. But these experiences have also provided a critical insider-outsider perspective from which he could address the intricacies of transitional justice. Rey Rosa's writings question justice in wider, more abstract terms than is often the case in testimonial interrogations of post–peace accord contexts. The aesthetic origins of his thought provide more expansive views on justice as a quality of the Good Life, rather than merely a series of practices that measure up or fail to measure up to political and juridical standards. Henighan explains, for example, that "cruelty enters the Guatemalan's repertoire for understanding his own country as a literary mannerism; but after Rey Rosa's return to Guatemala, cruelty and violence become anchored in concrete historical circumstances, undergoing a gradual transformation from aesthetic trait to social criticism" (Henighan, this volume, 179). However, Henighan clarifies that "Rey Rosa's social concern in *Lo que soñó Sebastian* is, in fact, less strictly social than environmental…. His initial concern for the country – 'El país más hermoso, la gente más fea' (… the most beautiful country, the ugliest people) as he would later write of the Guatemala City upper classes – takes the form of seeking to preserve its abundant natural life" (187). In *Los sordos* and *La cola del dragón* Rey Rosa exposes the actions of the ugly elite to demonstrate the absurdities of historical clarification as reconciliation, and to explore the cultural dimensions of justice. Henighan closes this multidimensional politico-literary analysis with a remarkably simple conclusion:

"The failure to recognize the fact of genocide is the key to stagnation of both human and environmental justice in Guatemala" (205).

Chapter 9 is written by George Lovell of Queen's University, an international expert on Guatemalan historical geography. His journey through the news headlines during his 2015 visit to Guatemala reveals deep contradictions: a post-conflict society with a formal, UN-negotiated peace accord that exists in parallel with increasingly rampant impunity and corruption; a demilitarized national regime with seemingly unlimited U.S. military support; an agrarian society that is a robust producer of foodstuffs for export while it suffers from increasing malnutrition and food insecurity; an abundance of natural resources that are exploited by foreign companies with little to no domestic return; and a pattern of presidents and other political officials who campaign on platforms of integrity and fairness in the midst of their own outrageous scandals. If justice is conceived as any reliable system, set of institutions, practices, or outcomes, then there is little hope to glean from the narrative threads that are daily woven through the post-war social fabric. As Lovell so aptly concludes, "Names are the consequences of things" (Lovell, this volume, 222). Guatemala has been labelled genocidal because, according to the UN, the internal armed conflict resulted in 200,000 dead and tens of thousands disappeared; most of the victims (83 per cent, according to the CEH; UN 1999, 102) were Indigenous persons massacred or forcibly disappeared by the army. The country is cast as corrupt because it is plagued by unimaginable scandals, unequal because virtually every economic and social indicator reveals this to be the case, a failed state because its institutions are weak and ineffective, and exploited because all of these elements of its character – genocidal, corrupt, unequal, failed – facilitate foreign commercial interests.

In the conclusion, chapter 10, we return to discussions of transitional, transnational, and distributive justice. As demonstrated throughout the book, the struggle for justice in Guatemala contains elements of all three types. The post–peace accords era is imbued with continued attempts at reconciliation in a divided country. These transitional attempts are both frustrated and facilitated by transnational actors. Further, endemic problems of distributional inequity seem to invite political action that maintains inequity, at the same time that those problems demand redress. Given the challenges faced by those pursuing justice, it is difficult to remain hopeful that the next attempt, or initiative, or decision, will bring about enough change to advance the cause. There are serious structural constraints to any form of progressive action. Political and judicial institutions are plagued by scandals, poverty and inequality are persistent, there are deep and racialized social divisions, and violence is pervasive at all levels. However, in spite of these constraints, there are many examples of the exercise of agency, individuals' ability to negotiate oppressive structural constraints, in order to effect change. The conclusion revisits the ways in which the various chapters demonstrate this point. The contributions by activists and practitioners – Magalí Rey Rosa, Helen Mack Chang, Claudia Paz y Paz – provide evidence of the lived experiences of this individual agency, and the contributions by academics – Catherine Nolin, Kalowatie Deonandan and Rebecca Tatham, Stephen Henighan, and Rita Palacios – document other experiences of social change and help to interpret the meaning of political, juridical, and cultural initiatives in the service of justice. In the conclusion we also discuss the role of the capital city in framing racial injustice and consider recent cultural expressions of resistance and perseverance by a new, post–peace generation in Guatemala.

These expressions of a new generation, in addition to the continued struggles documented in this book, suggest an evolution and indicate that hope may still be possible. Recent headlines are no more hopeful than those that Lovell reviews in chapter 9. President Morales is a polarizing figure who has great ambitions. He declared, "Mi aspiración más alta es que dentro de 100 años en las aulas escolares se enseñe que Jimmy Morales fue el mejor presidente de la historia guatemalteca" (Oliva Corado 2015; My greatest ambition is that 100 years from now it will be taught in schools that Jimmy Morales was the greatest president in Guatemalan history). His landslide victory (72 per cent support in the run-off ballot) is an indication that the status quo was no longer an option, and his anti-establishment profile and his bluster resulted in the press labelling him "Guatemala's answer to Donald Trump" (Lemos 2016). He has also recently signalled movement towards a more insular Guatemala that will be free from foreign political interference, which includes prohibition of observers at high-profile human rights violations and corruption trials (Pocasangre and Contreras 2016). This, however, is not realistic. Guatemala has an international audience – for its newspapers, its artists, and its progress towards peace and reconciliation. The international audience also has a responsibility to concern itself with progress towards justice in Guatemala. As the contributors to this volume demonstrate, Guatemala's transitional justice project is also transnational, as many of the enduring harms of the past are linked to American and Canadian corporate and political interests. Justice, therefore, must be sought in multiple venues and at multiple scales, and through both formal institutional channels and informal cultural progress. In fact, the latter – cultural expressions of justice and injustice – might be more important than the former – formal mechanisms such as trials and truth commissions. The two sets of formal and informal processes might also enable each other. Matthew Mullen makes a case for the primacy of attention to the cultural elements of injustice and its redress. He argues that "structural and cultural violence makes mass atrocities possible. This structural and cultural violence remains long after the atrocities are over, leaving violent systems intact. In other words, while these societies are in 'transition,' they continue to be subjected to violent systems and social environments. Nonetheless, this structural and cultural violence, which manifests as systemic vulnerability and dehumanization, has remained a peripheral focus in transitional justice efforts" (2015, 463). In the next chapter, Catherine Nolin picks up this line of argument and discusses the significance of structural violence in Guatemala and its consequences for human and environmental justice.

REFERENCES

Arriaza, Laura, and Naomi Roht-Arriaza. 2008. "Social Reconstruction as a Local Process." *International Journal of Transitional Justice* 2 (2): 152–72. https://doi.org/10.1093/ijtj/ijn010.
Balcárcel, Pep. 2015. "Scandal-Ridden Baldetti Resigns as Guatemalan VP." *PanAm Post*, 12 May. https://panampost.com/pep-balcarcel/2015/05/12/scandal-ridden-baldetti-resigns-as-guatemalan-vp.
Barillas, Byron. 2016. "Víctimas de Chixoy no ven voluntad del Gobierno en cumplir con indemnización." *El Siglo* 21, 22 February. http://www.website.adivima.org.gt/wp-content/uploads/2016/02/Chixoy.pdf.
Beaudoin, Sophie. 2015. "New Corruption Scandal Plunges Guatemala's Government Further Into Crisis." *International Justice Monitor*, 26 May. https://www.ijmonitor.org/2015/05/new-corruption-scandal-plunge-guatemalas-government-further-into-crisis/.
Burgos-Debray, Elisabeth, ed. (1985) 2010. *I, Rigoberta Menchú: An Indian Woman in Guatemala*. Translated by Ann Wright. 2nd ed. London: Verso.
Carey, David Jr. 2013. *I Ask for Justice: Maya Women, Dictators, and Crime in Guatemala, 1898–1944*. Austin: University of Texas Press.
Chen Osorio, Carlos. 2009. *Historias de Lucha y de Esperanza*. Rabinal, Baja Verapaz: Asociación para el Desarrollo Integral de las Víctimas en las Verapaces Maya Achí.

Collins, Cath. 2010. *Post-Transitional Justice: Human Rights Trials in Chile and El Salvador*. University Park, PA: Penn State University Press.

Crosby, Alison, and M. Brinton Lykes. 2011. "Mayan Women Survivors Speak: The Gendered Relations of Truth Telling in Postwar Guatemala." *International Journal of Transitional Justice* 5 (3): 456–76. https://doi.org/10.1093/ijtj/ijr017.

Cullather, Nick. 2011. "Operation PBSUCCESS." In *The Guatemala Reader: History, Culture, Politics*, edited by Greg Grandin, Deborah T. Levenson, and Elizabeth Oglesby, 230–7. Durham, NC: Duke University Press. https://doi.org/10.1215/9780822394679-047.

Daley, Suzanne. 2016. "Guatemalan Women's Claims Put Focus on Canadian Firms' Conduct Abroad." *New York Times*, 2 April.

Dearden, Nick. 2012. "Guatemala's Chixoy Dam: Where Development and Terror Intersect." *Guardian*, 10 December. https://www.theguardian.com/global-development/poverty-matters/2012/dec/10/guatemala-chixoy-dam-development-terror.

Democracy Now. 2016. "Guatemala: U.S.-Backed Former Dictator Ríos Montt to Face Genocide Trial." 18 November. https://www.democracynow.org/2016/11/18/headlines/guatemala_us_backed_former_dictator_rios_montt_to_face_genocide_trial.

Duggan, Colleen, Claudia, Paz y Paz Bailey, and Julie Guillerot. 2008. "Reparations for Sexual and Reproductive Violence: Prospects for Achieving Gender Justice in Guatemala and Peru." *International Journal of Transitional Justice* 2 (2): 192–213. https://doi.org/10.1093/ijtj/ijn013.

Elías, José. 2016. "El Gobierno de Jimmy Morales se estrena con una crisis militar: El relevo del jefe del Estado Mayor de la Defensa saca a luz las profundas diferencias en la cúpula militar de Guatemala." *El País* (Madrid), 23 January. https://internacional.elpais.com/internacional/2016/01/23/america/1453576896_576039.html.

Grandin, Greg, Deborah T. Levenson, and Elizabeth Oglesby, eds. 2011. *The Guatemala Reader: History, Culture, Politics*. Durham, NC: Duke University Press. https://doi.org/10.1215/9780822394679.

Imai, Shin, Bernadette Maheandiran, and Valerie Crystal. 2014. "Access to Justice and Corporate Accountability: A Legal Case Study of HudBay in Guatemala." *Canadian Journal of Development Studies* 35 (2): 285–303. https://doi.org/10.1080/02255189.2014.908274.

INDE. 2015. "INDE lleva energía a Cubulco, Baja Verapaz," 4 March. http://www.inde.gob.gt/.

Isaacs, Anita. 2010. "At War with the Past? The Politics of Truth Seeking in Guatemala." *International Journal of Transitional Justice* 4 (2): 251–74. https://doi.org/10.1093/ijtj/ijq005.

Johnson, Candace. 2016. "Conceptualizing Transnational Civil Society in Guatemala." In *Re-imagining Community and Civil Society in Latin America and the Caribbean,* edited by Roberta Rice and Gordana Yovanovich. New York: Routledge.

Klippensteins. 2016. "Choc v. HudBay Minerals Inc. & Caal v. HudBay Minerals Inc." http://www.chocversushudbay.com/history-of-the-mine.

Lemos, Eirini. 2016. "Jimmy Morales: Guatemala's Clown Who Went on to Become President." *Telegraph* (London), 15 January. http://www.telegraph.co.uk/news/worldnews/centralamericaandthecaribbean/guatemala/12100871/Jimmy-Morales-Guatemalas-clown-who-went-on-to-become-president.html.

Logan, Nick. 2014. "Canadian Mining Companies under Fire for Latin America Operations." *Global News Online,* 14 June.

Malkin, Elisabeth, and Nic Wirtz. 2015. "Jimmy Morales Is Elected New President in Guatemala." *New York Times,* 25 October. https://www.nytimes.com/2015/10/26/world/americas/former-tv-comedian-jimmy-morales-seems-set-to-be-elected-guatemalas-president.html.

Martínez Barahona, Elena, Martha Liliana Gutiérrez Salazar, and Liliana Rincón Fonseca. 2012. "Impunidad en El Salvador y Guatemala: 'de la locura a la esperanza: ¿nunca más?'" *América Latina Hoy* 61:101–36.

May, Rachel. 2013. "Truth and Truth Commissions in Latin America." *Investigación & Desarrollo* 21 (2): 494–512.

Miller, David. 1995. *On Nationality.* Oxford: Oxford University Press.

Mullen, Matthew. 2015. "Reassessing the Focus of Transitional Justice: The Need to Move Structural and Cultural Violence to the Centre." *Cambridge Review of International Affairs* 28 (3): 462–79. https://doi.org/10.1080/095575 71.2012.734778.

Nolin Hanlon, Catherine, and Finola Shankar. 2000. "Gendered Spaces of Terror and Assault: The Testimonio of REMHI and the Commission for Historical Clarification in Guatemala." *Gender, Place and Culture* 7 (3): 265–86. https://doi.org/10.1080/713668875.

Oettler, Anika. 2006. "Encounters with History: Dealing with the 'Present Past' in Guatemala." *Revista Europea de Estudios Latinoamericanos y del Caribe* 81 (81): 3–19. https://doi.org/10.18352/erlacs.9645.

Oliva Corado, Ilka. 2015. "Jimmy Morales: el descarado que niega el genocidio en Guatemala." *TeleSur* Blog, 26 October. https://www.telesurtv.net/bloggers/Jimmy-Morales-el-descarado-que-niega-el-genocidio-en-Guatemala-20151026-0001.html.

Ontario Superior Court of Justice. 2012. Between Margarita Caal Caal, Rosa Elbira Coc Ich, Olivia Asig Xol, Amalía Cac Tiul, Lucia Caal Chún, Luisa

Caal Chún, Carmelina Caal Ical, Irma Yolanda Choc Cac, Elvira Choc Chub, Elena Choc Quib, and Irma Yolanda Choc Quib (Plaintiff) and Hudbay Minerals Inc. and HMI Nickel Inc. (Defendant). Amended Statement of Claim, 6 February. Court file no. CV-11-423077.

Pocasangre, H., and G. Contreras. 2016. "Jimmy Morales vuelve a señalar injerencia." *Prensa Libre*, 16 March. http://www.prensalibre.com/ guatemala/politica/lo-que-se-ve-no-se-pregunta-dice-morales-de-la-injerencia-extranjera.

Rawls, John. 1999. *The Law of Peoples*. Cambridge, MA: Harvard University Press.

Reynolds, Louisa. 2015. "Guatemalans take to the streets again as new corruption scandal rocks the Pérez Molina administration." *The Tico Times*, 24 May. http://www.ticotimes.net/2015/05/24/guatemalans-take-to-the-streets-again-as-new-corruption-scandal-rocks-the-perez-molina-administration Accessed 3 May 2016.

Rights Action. 2005. "Investigación Social de Comunidades Afectadas por la Represa Chixoy / Social Investigation of the Communities Affected by the Chixoy Dam, Guatemala." March.

Rodríguez, James. 2013a. "Genocide trial begins in Guatemala." MiMundo. org, 19 March. http://blog.mimundo.org/2013/03/2013-03-19-genocide-trial-begins-in-guatemala/.

Rodríguez, Manuel. 2015. "Baldetti suspende plan para 'limpiar' lago de Amatitlán." *La Hora*, 30 March. https://web.archive.org/ web/20150423211047/http://lahora.gt/baldetti-suspende-plan-para-limpiar-lago-de-amatitlan/.

Roht-Arriaza, Naomi. 2001. "Making the State Do Justice: Transnational Prosecutions and International Support for Criminal Investigations in Post–Armed Conflict Guatemala." *Chicago Journal of International Law* 9 (1): 79–106.

Russell, Grahame. 2014. "U.S. Government Holding World Bank and IADB Accountable to Ensure Reparations for Chixoy Dam Victims in Guatemala." *Upside Down World*, 21 January. http://upsidedownworld.org/archives/ guatemala/us-government-holding-world-bank-and-iadb-accountable-to-ensure-reparations-for-chixoy-dam-victims-in-guatemala/.

Schlesinger, Stephen, and Stephen Kinzer. 1982. *Bitter Fruit: The Untold Story of the American Coup in Guatemala*. New York: Doubleday.

Skaar, Elin. 2011. *Judicial Independence and Human Rights in Latin America: Violations, Politics, and Prosecution*. London: Palgrave Macmillan. https:// doi.org/10.1057/9780230117693.

Tecú Osorio, Jesús. 2006. *Memoria de las Masacres de Río Negro*. Río Negro and Rabinal: Fundación Nueva Esperanza and ICB Nueva Esperanza.

Viaene, Lieselotte. 2010. "Life Is Priceless: Mayan Q'eqchi' Voices on the Guatemalan National Reparations Program." *International Journal of Transitional Justice* 4 (1): 4–25. https://doi.org/10.1093/ijtj/ijp024.

Weld, Kirsten. 2014. *Paper Cadavers: The Archives of Dictatorship in Guatemala*. Durham, NC: Duke University Press. https://doi.org/10.1215/9780822376583.

World Bank. 2016a. *Country Overview: Guatemala*. http://www.worldbank.org/en/country/guatemala/overview.

World Bank. 2016b. *World Development Indicators: Guatemala*. https://data.worldbank.org/country/guatemala.

Yates, Pamela, dir. 2011. *Granito: How to Nail a Dictator*. Skylight Films.

Young, Iris Marion. 2011. *Responsibility for Justice*. New York: Oxford University Press.

2 Memory, Truth, Justice: The Crisis of the Living in the Search for Guatemala's Dead and Disappeared

CATHERINE NOLIN

Introduction

In Central America broadly, and Guatemala specifically, the contemporary and overlapping crises of resource extraction, export-driven crop production, gang violence, and survival migration are devastating legacies of the region's internal armed conflicts that culminated, in Guatemala, in state-directed terror, massacres, and forced disappearances in the countryside. The physical erasure of Indigenous Maya communities and progressive, mostly urban Ladinos is central to these crises – and well documented in both the REMHI report *Guatemala: Nunca Más* (1998), and *Memoria del Silencio*, the report of the United Nations–sanctioned CEH (1999). These reports emphasize an almost forgotten component of contemporary studies of neoliberal development and escalating violence. Drawing on collaborative research and activism over many years, I explore some of the human dimensions of structural violence, and the accompanying suffering that delayed justice, violence, and impunity creates among surviving family members. I draw on three illustrative cases in the contemporary search for justice, truth, memory, and historical clarification, particularly among the search for the disappeared. I write from a place of witnessing. The three cases explore the search for the disappeared at La Verbena (Guatemala City's general cemetery), the Comalapa former military installation, and the former military garrison where the Comando Regional de Entrenamiento de Operaciones de Paz (CREOMPAZ; Regional Centre for Training in Peacekeeping) is now based. These cases highlight the creative ways that organizations such as the Fundación de Antropología Forense de Guatemala (FAFG; Guatemalan Forensic Anthropology Foundation) and surviving family members work to document the truth of their suffering and seek justice.

Structural Violence

Structural violence, Paul Farmer explains, is evident when social forces are at work that structure the risk for most forms of extreme suffering such as hunger, rape, and torture. But why, Farmer asks, are certain kinds of suffering so readily observable and yet "structural violence all too frequently defeats those who would describe it?" (2010, 335). He offers three reasons: (1) exoticization: the suffering of those remote (geographically and socially) from ourselves is somehow less affecting; (2) the sheer weight of suffering makes it more difficult to render; and (3) the dynamics and distribution of suffering are poorly understood. Farmer reminds us that historically given and often economically driven processes structure all kinds of social suffering due to structural violence; therefore, we need both geographically wide and historically deep analyses of violence in the community. Who will suffer abuse and who will be shielded from harm? Who benefits from neoliberal development and who suffers? Who gains from the suffering of others? Who is hurt? Farmer reminds us that, of course, the poor and Indigenous are most likely to suffer and are the least likely to have their suffering noticed. The unequal power we witness in Guatemala – with a set of military and elite alliances that, by all measures, have repeatedly and ferociously targeted a largely defenceless Indigenous and progressive Ladino majority population – generates brutality that is both obvious and quiet on the world stage.

This brutality of structural violence, geographer James Tyner (2012, 10) argues, is often shaped, reinforced, and driven by state policies and practices that work to institutionally legalize violence. Therefore, over time, violence is seen as instrumental and rationalized by the state for nation building and maintenance of the status quo. Tyner articulates in his work in Cambodia how a geographical perspective on mass violence must highlight how this form of violence, in the imagination of the perpetrators, is viewed as an effective and legitimate form of state building. State-directed violence is part of the making of the state – not the decomposition of the state, as many posit, and is clearly demonstrated in the Guatemalan case (Schirmer 1998).

The most recent form of state-orchestrated structural violence, on a massive scale, involves the violent transformation of land to "property," through the dispossession of Indigenous and poor non-Indigenous rural people. Contemporary neoliberalized development strategies involve, for example, the granting of mineral and property rights without consultation with, or consent of, affected communities. "Land fits within neoliberal economic and political discourse," Simon Granovsky-Larsen notes, "as a purely economic resource independent and devoid of social relations" (2013, 361), which encourages the acquisition of land for large-scale productive and extractive projects. McChesney sums up the impact of neoliberalism: "Neoliberalism is the defining political economic paradigm of our time – it refers to the policies and processes whereby a relative handful of private interests are permitted to control as much as possible of social life in order to maximize their personal profit" (1998, 7).

Culture, identity, and the environment have emerged as key neoliberal frontiers, according to Perreault and Martin (2005, 191), and these frontiers are linked by transnational practices. Taken together, these political, economic, and cultural processes are producing new geographies of neoliberalism. Neoliberalism is both a local and a global process often driving changes at a global or national level that embed themselves in the local landscape. This concept is central to mining projects in Guatemala where mining companies are guided by global and national policies, but the strongest impact of mining practices, as will be discussed by Magalí Rey Rosa in chapter 4, are felt in the local communities and environments.

Long-term Consequences: History in Extremis

I try to grapple with the long-term consequences of structural violence in Guatemala – the continuity and change evident over time – to explore how the search for memory, truth, and justice come together at this moment. To frame the long-term consequences of state-directed violence, Greg Grandin suggests that we not forget that it is situated in "history in extremis." He states, "The Guatemalan civil war in all of its cruelty could understandably be considered history in extremis – singular in its viciousness and devastation except that it so closely parallels and even propels much of the history of Latin America in the second half of the 20th century" (2004, 4).

All of the chapters in this book, in some form or another, document the legacy of military interventions, human rights abuses, and economic instability that followed the end of the "years of spring" with the coup against democratically elected President Jacobo Árbenz Guzmán in 1954 and the run of military dictatorships and military democracies[1] that followed. In 2011, in what the *New York Times* described as a "muted ceremony" at Guatemala City's National Palace, President Álvaro Colom told Árbenz's son Juan Jacobo, "That day [the coup] changed Guatemala and we have not recuperated from it yet. It was a crime to Guatemalan society and it was an act of aggression to a government starting its democratic spring" (Imison 2011).

Guatemalan Crises Revisited

Documentary photographer James Rodríguez has worked for years to document the search for memory, truth, and justice in Guatemala and situates his documentation and analysis in historical context: "People think the war ended. But the effects of the war are so current. To me, that has been a massive push factor in the whole immigration issue ... people talk about violence and gangs, but nobody connects it to the war, which they should" (cited in Gonzalez 2015).

Contemporary Guatemalan issues cannot begin to subside or to reverse until Guatemalans heal from the shattering of their communities, until justice is evident, until the bones of their lost loved ones are returned to them (Nolin Hanlon and Shankar 2000, 265–86). Surviving family members want the return of the more than 200,000 murdered and more than 45,000 "disappeared": those people strategically removed from society or forcibly disappeared for their ideals, intellect, political stance, and vision for a better Guatemala (Rodríguez 2013a).

Along with James Rodríguez, Grahame Russell (Rights Action), Erica Henderson, and Fredy Peccerelli (both of FAFG), I embarked on a project to interview families of the disappeared to understand what they mean when they say they are searching for justice. We are also trying to understand, geographically, the military strategy of disappearances. And we want to document visually the very human experience of living with traumatic memories, seeking truth through exhumations of mass graves and military and police records, and pursuing justice in a variety of ways. We are calling this the "crisis of the living." In their report on the global homicide rate, the United Nations argues that "the connection between violence, security, and development, within the broader context of the rule of law, is an important factor to be considered" (UNODC 2013a, 11). From our perspective, it is a crisis to be living in a world of impunity, where almost no one has been held accountable for the massive crimes of the past that perpetuate more crimes and violations in the present.

Jump off the Deep End

Jump off the deep end. That is what we do, together, when we step into the centre of death and destruction and see it through the eyes and words of people who are working to shed light on both the darkness of continued violence and the light: the many moments and people who love each other, work together, change and heal together. This lightness is sometimes hard to see or believe when situated in a city with one of the highest murder rates not just in the Americas but also in the entire world: more than 6,000 murders annually in recent years (International Crisis Group 2014). That is about 116 murders per 100,000 people in the capital (UNODC 2013b), while in Canada the murder rate is 1.45 murders per 100,000 (Statistics Canada 2015). But the point of our scholarship and activism is to see the light within the darkness: to witness real change and courage amid the endemic structural violence.

Through this research, we attempt to understand Guatemala in the context of North-South connections, through the scales of accountability (from local through to the North and South), and the necessity to bring to bear a historical perspective that asks, What is our responsibility in the North in creating and understanding Guatemala as an exploited and abused country rather than a "developing country"?

A stroll through Zone 1, the historical centre of Guatemala City, in May 2010, into and through the Parque Central (Central Park), past the National Cathedral, the National Palace, and into Zone 2 brought us to the office of the Asociación Familiares de Detenidos-Desaparecidos de Guatemala (FAMDEGUA; Association of Relatives of the Detained and Disappeared in Guatemala) to meet with Sonia Serrano. There is no sign on the door, but a beautiful mural reading "Memoria / Verdad / Justicia" (Memory/Truth/Justice) leads the way to a place of remembrance and struggle for justice for the thousands upon thousands of "disappeared" family members from the internal armed conflict of the 1970s, 1980s, and into the early 1990s. Among the many black-and-white photos of the disappeared on the walls, Sonia pointed to her dad, Jorge Luis Serrano. He was one of the twenty-seven members of the Centro Nacional de Trabajadores (CNT; National Workers' Union) forcibly disappeared on 21 June 1980 from the CNT's headquarters and never seen again. Sonia was just four years old at the time and laments the fact that she never really knew her father. It is coming up to the thirty-year anniversary, she reminds me. But here she is working with one of the most courageous organizations in Guatemala founded back in the early 1980s, to find him and the truth of his death.

When I reflect on this one day, an accumulation of sadness, fear, deep compassion, empathy, and more shaped our thoughts. These overwhelming feelings seemed to pour out of the small group collected together in the evening while meeting with Mischa Prince to screen segments of his forthcoming documentary film *Still Alive* based on his more than twenty years of working in Guatemala. We sat in my room – room 9 – of the Spring Hotel in the dark. All of us, with Grahame, Mischa, lawyer Fernando López, who spoke briefly with the group earlier in the day, as well as Carlos Chen, who is a Río Negro massacre survivor and community leader in the fight for compensation and reparations for the thirty-three affected communities of the Chixoy dam project of the late 1970s and early 1980s (Einbinder 2014; Johnston 2010).

Carlos happened to be in the capital that day in preparation for the formal signing of an agreement between the affected communities and the Guatemalan government (after President Álvaro Colom, just one month earlier, officially "accepted responsibility for the harms and violations caused by the project and agreed to pay reparations") (Russell 2014). Years upon years of negotiations and struggle led by the survivors was about to culminate in an official agreement to be signed in the National Theatre in the capital of the country. We planned to attend. Hundreds of people also planned to make their way from the countryside by bus the next day in order to witness this momentous signing.

It was all for naught. Carlos arrived at the Spring Hotel and sat quietly at the entrance while waiting for his room. The government backed out, he told me, in the final hour. He had lots of calls to make and receive.

"No, don't send the buses. No, no agreement yet. No, no financial compensation yet."

No land, no plans, no. It fell through. Again. Kicked in the teeth one more time. And he, not the government, was the bearer of the bad news.

In the late afternoon, Carlos met with us to recount, detail by detail, disappointment by disappointment, lie by lie, the ways in which he and others like Juan de Díos have fought for their people, for compensation and reparations after suffering the massacres and the subsequent decades of poverty, misery, displacement, ill health, and psychological trauma of injustice (COCAHICH 2010). The presentation was technical, recounting facts and details to give us a sense of the enormity of the communities' reactions to this latest disappointment.

We all held it together until we returned to the Spring Hotel from dinner to watch several segments chosen by filmmaker Mischa Prince from *Still Alive*. The film is Mischa's years'-long journey to tell the story of two young Guatemalan men who are part of the theatre group Caja Lúdica (Magic Box): Fu, a former gang member, and Juan Carlos, whose father belonged to a death squad.

Before screening the film segments, Mischa recounted the beautiful and brutal story of Tiburcio that will stay with me forever when grappling with the dark and light of life after violence. During his work with Comunidades de Población en Resistencia (CPRs; Communities of Population in Resistance), who hid from the military in the mountains for years, Mischa met and listened for more than eight hours to Tiburcio's testimonio.[2] Tiburcio was captured by the military and held for one and a half years in the most horrendous and terrifying conditions, thrown into a room of half dead, dead, and decomposing bodies. Night after night he was made to sit on a chair in this room and in the morning the soldiers would ask him, "Tiburcio, did you dream of tigers? Did you dream of lions? Did you dream of vultures?" And each time, Tiburcio would respond, "No, I dreamed of angels. No, I dreamed of angels. No, I dreamed of angels."

After a long series of events, Tiburcio was able to ask to leave this place and collect his family. God only knew where they were after all this time. He began his search. More than half starved, he walked three days and three nights into the mountains, following only the sound of the cockerel. Every day he followed the call of the cockerel until he finally came to a small house and knocked on the door. His sister answered: "Is that you, Tiburcio? Are you the devil in the shape of my brother? Did you bring the army?"

"No," he said, "it is finally me, Tiburcio."

Through that terrible and beautiful testimonio – one of thousands upon thousands – Mischa Prince revealed both the dark and lightness in Guatemala. The atrocious, vicious actions of the state-directed terror, accompanied by the lightness of finding a loved one. Moving forward together. Reweaving the social fabric.

Carlos watched the film with us. He watched Mischa filming in his home community, in the Community Genocide Museum, with the black-and-white photos of his loved ones, his friends, community members, his wife. He watched as Mischa met with a former death squad member, as he recounted what he did, what they all did, to people like Carlos, his pregnant wife, and his children. We could barely breathe knowing that Carlos was sharing this experience with us, and only he could know what it really meant, in a deep way that we could never experience. Fernando López, too, watched with us, sitting beside Carlos, listening as Carlos pointed out loved ones on the screen. Fernando knows these stories well, as he is the lawyer charged with moving ten paradigmatic criminal cases through the broken justice system.

The lights came on.

"No, not the lights!" people called, perhaps not ready to share tears and red eyes.

The first to speak was Fernando. "Well," he said quietly, "what can I say? I cried through the whole thing. I think this is perhaps the finest documentary film ever made on Guatemala's story."

I think it was knowing that Fernando, too, cried that made it all right for the others in the room. Why hide the tears? If we cannot cry about genocide, and cruelty beyond belief, and for the youth of today who must somehow navigate extreme social violence, domestic violence, a judicial system that only abuses or ignores them, a national government that tells them they are vile and unworthy of care, what can we cry about? A young man in the film attending the funeral of a good friend struggled with the massive emotions of wasted lives. "Here, we have never learned to cry. I feel like I just want to shoot someone."

As we will find out together over these years, there is no end of causes for tears. The Guatemalan government finally made first payments from the Chixoy Dam Reparations Plan in October 2015 – only a small portion of the 1.2 billion quetzales ($154,000,000) owed. As Monti Aguirre points out, "Even if the entire amount of reparation funds owed … is paid out one day to the thousands of Chixoy Dam victims, it cannot compensate for all who were killed, for the homes and communities destroyed forever, for the generations of lost livelihood" (Aguirre 2015).

And it also starts at La Verbena Cemetery, Guatemala City's general cemetery, which is the site of the FAFG's ongoing exhumation (Smith 2012). Fredy Peccerelli, the FAFG's executive director, has invited us several times to visit the forensic lab and join the team at La Verbena to witness their most significant work, the beautiful work of recovering the bones of the disappeared: the university students, faculty members, labour lawyers, union leaders, resistance supporters, and so on who make up the 45,000 people torn from their families and networks of association and never seen again (Kobrak 1999, 5). In both urban and rural terrorized spaces, Guatemalans witnessed government-sanctioned violence, the killings and disappearances of tens of thousands of people, and in the final stages, the massive displacement of more than a million Indigenous and non-Indigenous people within the country, and at least a million people beyond the borders (Nolin 2006, 3). The FAFG are convinced that many of the disappeared are in the bone well in the middle of the cemetery. Disappeared in plain sight, in the centre of the city.

La Verbena: Crisis of the Living, Centre of Hope

In the early 2000s, Fredy Peccerelli and Dr Clyde Snow, the well-known grandfather of forensic anthropology, first hypothesized and later discovered that La Verbena was "hiding in plain sight" hundreds of people known to be disappeared from the internal armed conflict (Snow et al. 2008, 89). This case of political violence pushed the methodology of the FAFG to incorporate new methods for best results in identifying the disappeared. It was necessary to incorporate DNA analysis into the strategy, otherwise it would be impossible to identify anyone when the team did not know exactly who was placed in the bone wells and when.

One doctoral student accompanying me wrote a short reflection after her first experience at La Verbena.

> I peered down, way down, holding tightly to the yellow rope preventing myself from falling into the abyss – the pile of all of those bones. Of course the rope was there for safety, but as I peered down into the well, I felt it more of a lifeline. Oh God, keep me safe.
>
> When they asked me to help to clean the bones – femurs, jawbones, and skulls – I thought, "No, I don't want to touch such death, horror, pain." The truth of Guatemala's history in my hands. "Just sweep the dirt off like this …" Claudia Rivera explains while holding a large femur and a paintbrush. I held those bones bringing the truth of Guatemala's violent history too close for my comfort. (Bois 2010)

The space of La Verbena and its meaning are now transformed into a centre of hope for families in the search for the disappeared. After the investigation commenced on 7 October 2009, families began visiting the FAFG at La Verbena to provide testimonios and DNA samples with the hope that their loved ones might be recovered (Rodríguez 2010). The FAFG exhumed 15,557 skeletal remains from the five bone wells at La Verbena (FAFG

2015a). From the analysis thus far completed, eight victims of enforced disappearance are now identified from La Verbena, including nineteen-year old José Zenon Hernández Cuzanero, the eighth identification, who was exhumed on 5 October 2011 in the second bone well and listed as #112 in the Diario Militar (1999). Much analysis remains to be done; however, as DNA work is expensive, the FAFG is searching for way to fund the analysis of the samples that remain to be analysed to then be compared in the database against the approximately 13,000 family reference samples.

Comalapa and the Diario Militar: Crisis of the Living and Investigative Strategy

Another case, which employs the same strategy, is that of the Comalapa former military installation. But before we can examine the findings of the FAFG's Comalapa exhumation, we must return to the Death Squad Dossier/Diario Militar.

The Death Squad Dossier identifies 183 targeted victims of Guatemalan security forces from 1983 to 1985, many of whom were university students affiliated with an opposition group or members of the student executive body. Of the 183 victims, 13 women are listed with ties to opposition groups such as Partido Guatemalteco de Trabajo (PGT; Guatemalan Workers' Party) (7), Organización Revolucionaria del Pueblo en Armas (ORPA; Revolutionary Organization of the People in Arms) (4), and Fuerzas Armadas Rebeldes (Rebel Armed Forces) (2), while no group affiliation is noted for 11 women (e.g., they lived in raided safe houses). This chilling document, the first of its kind to come to light, includes the names, photographs, and summary information about each victim, including opposition group membership, the date of kidnap, and how the kidnap operation functioned along with the final outcome (Doyle 2000). For some, this meant capture and torture for a month or more before release or death, which is coldly noted by the number 300.

The significance of this document is highlighted by Kate Doyle, senior analyst and director of the Guatemala Documentation Project at the National Security Archive:

> The State of Guatemala has systematically hidden the information in its power about the internal armed conflict. The Guatemalan Army, the Police and the intelligence services are intrinsically opaque, secretive and closed institutions, and it has been almost impossible to gain access to their records. This policy of silence has survived the peace accords; it has

survived the Historical Clarification Commission; and it continues today – despite the discovery of archives, the exhumations of clandestine cemeteries, the criminal convictions of perpetrators of human rights violations, and the unceasing demand for information by families of the disappeared ... the Historical Clarification Commission (Comisión para el Esclarecimiento Histórico-CEH) ... from its earliest days tried to obtain information from State security agencies, but without success. During the very same week in which the CEH report was made public, in February 1999, the Military Logbook appeared ... The document was turned over to me and I made it public three months later in a press conference in Washington. It was exactly the kind of information sought repeatedly by the CEH, without success. (Doyle 2012)

In 2003 the FAFG obtained access to the former military installation at Comalapa, where the team exhumed fifty-three graves and recovered 220 bodies. Concurrently, the family of Sergio Saúl Linares Morales visited the FAFG's exhumation site at the bone wells of La Verbena to provide their testimonio and DNA samples, wondering if their son was hidden among the thousands who met their end at La Verbena.

However, there was a DNA "hit" or match from the exhumation at the Comalapa military installation. Sergio is #74 in the Death Squad Dossier; his entry is accompanied by the 300 code, giving the date of his death as 29 March 1984.

After reviewing the Death Squad Dossier and completing their analysis of the Comalapa exhumation, the FAGF identified six individuals from the Diario Militar, all with the same "300" date of execution, and all were found in the same grave.

The death squads disappeared these six men on different dates, from different locations, but killed them all on the same day and buried them in the same grave in the Comalapa military base. The FAFG understands that this new information and evidence provides some insight into the strategy of enforced disappearance employed by the military. This set of identifications opens spaces and opportunities for justice that was not possible without the FAFG's forensic investigations.

CREOMPAZ and Evidence of "Largest Case of Forced Disappearance in Latin America"

The crisis of the living brought the FAFG and their investigative strategy to CREOMPAZ, the former Guatemalan Military Garrison No. 21

in Cobán, Alta Verapaz. The Guatemalan Military established Garrison No. 21 in 1971, five kilometres outside of Cobán, which is a major urban centre of the eastern highlands, and the CEH established it as one of the areas where enforced disappearance and the violence of the internal armed conflict hit hardest outside the capital. The Guatemalan Army has maintained and continues to maintain control of the grounds since its opening in 1971. In May 2004 the government deactivated the military base and recreated it as CREOMPAZ. On 27 February 2012 the FAFG began exhumations of mass graves within the CREOMPAZ grounds, under Public Ministry orders, and based on witness testimonios, which stated that burials were observed within the perimeter of the military garrison between 1980 and 1996. Three months later, the FAFG team met our University of Northern British Columbia (UNBC) / Rights Action delegation of undergraduate and graduate students accompanied by photographer James Rodríguez of MiMundo.org, at the Cobán CREOMPAZ excavation (Henderson and Nolin 2012a, 2012b). We obtained Public Ministry clearance for access to the former military garrison and only James Rodríguez was permitted to take photographs. Representatives of the attorney general and the Guatemalan Army's legal department monitor the site, as do military trainees.

The exhumation within CREOMPAZ was procedurally different from normal investigations. An assigned prosecutor and a military defence lawyer had to be present at all times to document activities. Access for relatives to the CREOMPAZ site, when granted, was very limited, unlike exhumations in the countryside, which are supported and mindfully observed by surviving family members.

The FAFG completed the investigation on 13 December 2013, including excavation of eighty-three graves and recovery of 535 individuals (FAFG 2015b, 1). The FAFG excavated twenty-six individual graves, forty-one mass graves, and seventeen graves with an established Minimum Number of Individuals (FAFG 2015b). In addition, 389 artefacts of torture were recovered with 289 skeletal remains, including artefacts used as blindfolds, tied necks, wrists, and ankles, as well as shirts covering heads, necks, and mouths (Henderson and Nolin 2012b; Henderson, Nolin, and Peccerelli 2014). Of the 360 skeletons analysed by April 2015, 139 were blindfolded, 34 had ties around their thorax, 11 had ties associated with upper and lower extremities, 60 with hands tied together, 12 with feet tied together, and 5 with the waist and chest tied. All of this was found on the grounds of a military base that is now used as a UN training base to promote human rights; peacekeeping training takes place on top of mass graves. UNBC graduate student Erica Henderson reflected on this experience in her field journal:

> Without hesitation, Ramiro ripped open the individual's pants to expose the bones for further site photos. He asked me to do the same to the other skeletal remains we were preparing to exhume. Since the clothing had been buried since the 1980s, the seams were pretty fragile. It requires little effort, however it takes more strength to knowingly rip someone's last set of clothing. Very difficult considering family members might be holding onto the last images of their loved one before their disappearance wearing these clothes, this shirt, and those pants. The clothing that individuals were unjustly killed wearing. I ripped the other pant leg open, continuously repeating "I am sorry" in my mind and in my heart. (2012)

Seventy victims have been identified from CREOMPAZ, sixty-five males and five females (FAFG 2015b, 3), and fifty-two individuals have been returned to their families and reburied.

The identifications, based on place and date of disappearance, start to help us illustrate the geographical area covered by the military during the country's internal armed conflict. As well, disappearances from locations such as San Cristóbal Verapaz, Santa Cruz Verapaz, El Estor Izabal, Cahabon, Chisec, La Tinta, Rabinal, Panzos, San Juan Chamelco, and San Pedro Carcha could indicate that Military Garrison No. 21 functioned as a detention centre and unofficial burial place for detained and disappeared persons. Using the information gathered from the identifications, the FAFG is now targeting and strengthening its investigation in certain geographical areas to increase the chances of identifications, in addition to developing a solid scientific strategy to find the disappeared.

The seventy identifications bring to light the truth of two specific events during the internal armed conflict. On 2 June 1982 the military entered the community of Pambach, San Cristóbal Verapaz, Alta Verapaz, and, according to witness testimonio, marched eighty-six men away from the place for "military service." The FAFG exhumed sixty-four bodies in grave #17 and positively identified twenty-eight of those bodies to be from Pambach. All twenty-eight were disappeared on the same day and then dumped in one mass grave. One community, one killing event. The FAFG has confirmed the identity of the twenty-eight men from Pambach and returned them to the surviving families for reburial, beautifully documented by James Rodríguez (2013b).

The second event is the forced disappearance of women and children from Los Encuentros, Río Negro, Rabinal, Baja Verapaz. Testimonio tells us that on 14 May 1982 military soldiers surrounded Los Encuentros, killed approximately forty people, and forcibly disappeared fourteen women and forty children, taking them away in helicopters. Two women and one male child found at CREOMPAZ are now identified from grave FAFG 1433-XV among the bones (FAFG 2015b, 4). On 14 May 2014, on the thirty-second anniversary of the Los Encuentros massacre and disappearances, we witnessed a bit of justice: Grahame Russell, James Rodríguez, and I accompanied the family during the wake and reburial of their son Manuel Chen, who was three years old when the military took him away and killed him (Rodríguez 2014a, 2014b). During the same week as Manuel's wake, Guatemala's Congress approved a non-binding resolution that denies there was any attempt to commit genocide during the thirty-six-year internal armed conflict (*Cultural Survival Quarterly* 2014). "Who, then, killed Manuel?" asked his parents, Carmen Sánchez Chen and Bernardo Chen.

Against all odds and death threats, on 6 January 2016 Guatemala's Public Prosecutor's Office (Ministerio Público 2016a) issued arrest warrants and captured fourteen high-ranking former military officers, eleven with ties to Military Garrison No. 21 during the internal armed conflict (Castillo and Ríos 2016), and charged them with crimes against humanity involving massacres and disappearances of people by security forces under their command (De León 2016). "We are talking about one of the biggest cases of forced disappearance in Latin America," stated Guatemala's Attorney General Thelma Aldana (Ministerio Público 2016b) on the morning of the arrests.

The captured military officials include retired General Manuel Benedicto Lucas García (the brother of former president Fernando Romeo Lucas García, who ruled Guatemala between 1978 and 1982, during some of the worst of the violence), Manuel Antonio Callejas Callejas (director of the G2 Military Intelligence during the same period), and Francisco Luis Gordillo Martínez (who helped bring former dictator President General Efraín Ríos Montt to power from 1982 to 1983) (Prensa Communitaria 2016). Many are affiliated with the organization of retired military officers, Asociación de Veteranos Militares de Guatemala (AVEMILGUA; Association of Military Veterans of Guatemala), the power behind President Jimmy Morales, a comedian who was officially installed just one week after these historic arrests.

As Grahame Russell points out, "For years, the [Ríos Montt] genocide trial was basically the only – albeit extraordinary – case against a high-ranking military officer, for crimes against humanity. Now, in one day, crimes against humanity charges have been filed against 14 former high ranking military officers [eleven in relation to CREOMPAZ, three in relation to another case]" (teleSUR English 2016).

The two cases of the FAFG's positive identifications, from Pambach and Los Encuentros, now form a key part of these cases of crimes against humanity, planned, strategized, and implemented by high-ranking military officers who are alleged to have used Military Garrison No. 21 as their detention, torture, and killing centre.

Justice?

Can surviving family members expect more than the return of their loved ones' bones? Can the crisis of the living be transformed to something more? The CREOMPAZ case, among others, offers perhaps the best chance at recovery of memory, clarification of the truth of horrifying events, and legal justice. But what do family members mean by justice? What do family members and those involved in these processes mean when they state that they are seeking justice? When family members of disappeared loved ones are asked what justice means to them, several themes emerge (Henderson 2017) often in the same breath, illustrating the interconnectedness of all aspects of justice mentioned by relatives and those working with cases of enforced disappearance.

First, families spoke of *legal justice*, which involves prosecuting those who committed and planned these crimes, to ensure that these crimes will not occur again, and offers a challenge to deep impunity. Second, many interviewed family members indicated the importance of *change in the political power* in Guatemala. The political climate in which the elites (military, financial, religious) from the time of the internal armed conflict retain power makes it difficult to attain legal justice. A third theme was the crucial component of *acknowledgment*. Key to family members is that Guatemalan society and the state acknowledge that the forcibly disappeared were not criminals, that they did nothing wrong. Family members know that their lost loved ones did not deserve their fate, and they did not receive a due and just process once detained by state forces. Finally, interviewees spoke of the *needs of surviving family members to know*: family members want to know the truth, they want the remains of their lost loved ones back, and they want to bury their loved ones with dignity.

This spectrum of justice supports dignification, truth, and memory of the disappeared and their relatives. These points are attained by applying forensic science to cases that investigate the grave human rights violations from the conflict. The FAFG applies forensic science in the service of life and human rights. The pain of the living, of surviving family members, in the search for the disappeared for memory, truth, and justice is now coupled with the work of the FAFG.

But this hopeful search operates in a context of continued disappearances, continued impunity, with only a handful of high-profile cases such as CREOMPAZ shining a light into the darkness (attempts to use the legal system to end impunity are addressed by Helen Mack Chang and Claudia Paz y Paz respectively in chapters 5 and 6). The Grupo de Apoyo Mutuo (GAM; Mutual Support Group), a high-profile NGO of families of the disappeared, released findings on contemporary disappearances that cement our portrait of the crisis of the living. The issue of the disappeared is not confined to the internal armed conflict; the most recent data show a rapidly growing number of contemporary disappearances. Since 2003, the GAM documents more than 25,000 disappearances, with men and women disappearing in almost equal numbers, but with more women than men disappeared in recent years (GAM 2015). Disappearances, to many, are perhaps the cruellest crime, one that never ends; to always be nowhere. And "getting one's hopes up for tomorrow" is, according to Kirsten Weld, "a risky proposition" (2014, 208). But the search continues because it must.

Upon receiving the Puffin/Abraham Lincoln Brigade Archives Award, one of the world's largest monetary awards for human rights,[3] Fredy Peccerelli spoke truth to power in the face of threats and ongoing violations:

> I want to end by dedicating this award to the 201 victims of the Dos Erres Massacre, the 268 victims of the Plan de Sanchez Massacre, the 424 victims of the Cuarto Pueblo Massacre, the 177 women and children executed at the Río Negro Massacre, and to the victims of all the massacres in Guatemala. But not only to them. Also to Amancio Villatoro, Sergio Linares, Juan de Dios Samayoa, Hugo Navarro, Moises Saravia, and the other 178 victims that appear in the Military Diary, as well as the other 45,000 victims of enforced disappearance. My message to you is "We are looking for you and we will find you. Your stories will be told to every Guatemalan and to the world." For truth, justice, and for dignity. (Peccerelli 2012)

We will find you. We will tell your stories. We will all continue to work together to fight against structural violence, clarify the military strategy of massacres and disappearances, to honour the disappeared and their hopes for a better world.

NOTES

1 By "military democracies" I mean to flag the influential role that the military and former military officials continue to play at the highest levels of government and within political parties in Guatemala. Back in 1988, former army vice chief of staff and director of the Army General Staff (1982–3) Héctor Gramajo made clear "a civilian government … allowed the army to carry out much broader and more intensified operations because the legitimacy of the government, in contrast to the illegitimacy of the previous military governments, doesn't allow the insurgency to mobilize public opinion internationally" (Cruz Salazar 1988, cited in Schirmer 1998, 34).

2 I maintain that testimonio emerges from a research method used to bring previously silenced voices into research projects; when the researcher is in solidarity with the research community and their ongoing struggle. Social scientists must be flexible when conducting in-depth interviews and accessing testimonios so that the testimonio can capture the collective experiences of marginalized individuals or communities usually about the systems or situations that have historically suppressed their ability to speak. Testimonio empowers those affected by removing them from a position of powerlessness to a position of power, or from object to subject, through the act of sharing the oppressive or exploitative event (Nolin Hanlon and Shankar 2000, 268; Bois and Nolin 2011). This definition of testimonio differs from the literary use developed by John Beverley (1993, 70), which is employed by Henighan in chapter 8.

3 The Puffin/ALBA Award's $100,000 cash prize is granted annually by Abraham Lincoln Brigades Archive and the Puffin Foundation to honour the International Brigades and connect their inspiring legacy with contemporary causes. http://www.puffinfoundation.org/projects/puffin-alba-human-rights-award.html.

WORKS CONSULTED

Aguirre, Monti. 2015. "Chixoy Reparations at Last: Checks Are In." International Rivers, 20 October. https://www.internationalrivers.org/blogs/233/chixoy-reparations-at-last-checks-are-in-0.

Beverley, John. 1993. *Against Literature.* Minneapolis: University of Minnesota Press.

Bois, Claudette. 2010. "Dispatch #3 Day 2: *Mi nombre no es XX,* La Verbena, Guatemala City Cemetery." 12 May. https://www.facebook.com/notes/catherine-nolin-hanlon/guate-2010-dispatch-3-mi-nombre-no-es-xx/396295411929.

Bois, Claudette, and Catherine Nolin. 2011. *"Testimonio:* A Tool for Documenting the Lived Realities of Mining-Affected Maya-Q'eqchi' Communities in El Estor, Guatemala." Poster presentation, Canadian Association of Geographers, Annual Meeting, University of Calgary, 2 June.

Castillo, Sofía, and Gabriela Ríos. 2016. "Comprender el juicio a los militares en 110 segundos." *Nómada,* 18 January. https://www.youtube.com/watch?v=K3h1TyjFpT4.

Comisión para el Esclarecimiento Histórico (CEH). 1999. *Guatemala: Memoria del silencio (Tz'inil na 'tab'al).* Guatemala: Comisión para el Esclarecimiento Histórico (Spanish language). Guatemala, Guatemala: CEH.

Coordinadora de Comunidades Afectadas por la Construcción de la Hidroeléctrica Chixoy (COCAHICH). 2010. *Plan de reparación de daños y perjuicios sufridos por la comunidad afectadas por la construcción de la hidroeléctrica Chixoy.* Rio Negro, BV, Guatemala: ADIVIMA.

Cultural Survival Quarterly. 2014. "Congress in Guatemala Officially Denies Genocide." 14 May. https://www.culturalsurvival.org/news/congress-guatemala-officially-denies-genocide.

De León, Quimy. 2016. "Militares detenidos por violaciones a derechos humanos durante la Guerra." *Prensa Comunitaria Km. 169,* 6 January. https://comunitariapress.wordpress.com/2016/01/06/militares-detenidos-por-violaciones-a-derechos-humanos-durante-la-guerra/.

Diario Militar / Military Diary. 1999. *Guatemalan Military "Death Squad Dossier," 1983–1985* (a secret military intelligence dossier published by the National Security Archive in Washington, DC, in 1999. https://nsarchive2.gwu.edu//NSAEBB/NSAEBB15/dossier-color.pdf.

Doyle, Kate. 2000. "The Guatemalan Military: What the U.S. Files Reveal." 1 June. National Security Archive Electronic Briefing Book No. 32. Washington, DC: National Security Archive. https://nsarchive.gwu.edu/NSAEBB/NSAEBB32/index.html.

– 2012. "Update: The Guatemalan Death Squad Diary and the Right to Truth: National Security Archive Expert Testifies before International Court." 3 May 2012. National Security Archive Electronic Briefing Book No. 378. Washington, DC: National Security Archive. http://nsarchive.gwu.edu/NSAEBB/NSAEBB378/.

Einbinder, Nathan. 2014. "Río Negro Rebuild and Face the State." *NACLA Report on the Americas* 47 (2): 14–19. https://doi.org/10.1080/10714839.2014. 11721844.

Farmer, Paul. 2010. "On Suffering and Structural Violence: Social and Economic Rights in a Global Era (1996, 2003)." In *Partner to the Poor: A Paul Farmer Reader*, edited by Haun Saussy, 328–49. Berkeley: University of California Press.

– 2015a. "Comunicado de Prensa: Cementerio la Verbena," 5 January.

– 2015b. "Information Sheet for CREOMPAZ Investigation, Cobán, Alta Verapaz," 20 January.

Gonzalez, David. 2015. "Searching for Peace and Justice in Guatemala." *New York Times*, 15 April. https://lens.blogs.nytimes.com/2015/04/15/searching-for-peace-and-justice-in-guatemala/.

Grandin, Greg. 2004. *The Last Colonial Massacre: Latin America in the Cold War*. Chicago: University of Chicago Press. https://doi.org/10.7208/chicago/9780226306872.001.0001.

Granovsky-Larsen, Simon. 2013. "Between the Bullet and the Bank: Agrarian Conflict and Access to Land in Neoliberal Guatemala." *Journal of Peasant Studies* 40 (2): 325–50. https://doi.org/10.1080/03066150.2013. 777044.

Grupo de Apoyo Mutuo. 2015. "25,222 desapariciones en 12 años en Guatemala. Gráfica elaborada por el área de transparencia." GAM, Ministerio de Gobernación, 29 April. http://areadetransparencia.blogspot. ca/2015/04/25222-desapariciones-en-12-anos-en.html.

Henderson, Erica. 2017. "Seeking Justice in Guatemala: Dignifying the 'Disappeared' in a Context of Impunity." MA Interdisciplinary Studies, University of Northern British Columbia, Prince George, BC, Canada.

Henderson, Erica, and Catherine Nolin. 2012a. "448 bodies and Counting at the Cobán Former Military Garrison, Guatemala." *Rights Action Newsletter*, November.

– 2012b. "Covered Eyes, Hands Tied: Reflections and Exhumations at the CREOMPAZ Former Military Garrison, Cobán, Guatemala." Poster presentation, Conference of the Canadian Association for Physical Anthropology, University of Victoria, Victoria, BC, 4–7 November.

Henderson, Erica, Catherine Nolin, and Fredy Peccerelli. 2014. "Dignifying a Bare Life and Making Place through Exhumation: Cobán CREOMPAZ Former Military Garrison, Guatemala." *Journal of Latin American Geography* 13 (2): 97–116. https://doi.org/10.1353/lag.2014.0027.

Imison, Paul. 2011. "Justice and Jacobo Árbenz in Guatemala: 1954 Revisited." Upside Down World: Covering Politics and Activism in Latin America, 28

October. http://upsidedownworld.org/archives/guatemala/justice-and-jacobo-arbenz-in-guatemala-1954-revisited/.

International Crisis Group. 2014. "Corridor of Violence: The Guatemala-Honduras Border." *Crisis Group Latin America Report* 52, 4 June. Guatemala City. https://www.crisisgroup.org/en/regions/latin-america-caribbean/guatemala/052-corridor-of-violence-the-guatemala-honduras-border.aspx.

Johnston, Barbara Rose. 2010. "Chixoy Dam Legacies: The Struggle to Secure Reparation and the Right to Remedy in Guatemala." *Water Alternatives* 3 (2): 341–61.

Kobrak, Paul. 1999. *Organizing and Repression in the University of San Carlos, Guatemala, 1944 to 1996*. Washington, DC: American Association for the Advancement of Science and Centro Internacional para Investigaciones en Derechos Humanos.

McChesney, Robert W. 1998. "Introduction." In *Profit over People: Neoliberalism and Global Order*, edited by Noam Chomsky, 7–16. New York: Seven Stories.

Ministerio Público. 2016a. "Conferencia de Prensa de Ministerio Público: Investigación MP CASO Creompaz." https://www.youtube.com/watch?v=lPzNRlUFHyE.

– 2016b. "CREOMPAZ: 'Con este caso estamos hablando de uno de los mayores de América Latina de desapariciones forzadas,'" 6 January. https://www.mp.gob.gt/noticias/2016/01/06/creompaz-con-este-caso-estamos-hablando-de-uno-de-los-mayores-de-america-latina-de-desapariciones-forzadas/.

Nolin, Catherine. 2006. *Transnational Ruptures: Gender and Forced Migration*. Aldershot, UK: Ashgate.

Nolin Hanlon, Catherine, and Finola Shankar. 2000. "Gendered Spaces of Terror and Assault: The Testimonio of REMHI and the Commission for Historical Clarification in Guatemala." *Gender, Place and Culture* 7 (3): 265–86. https://doi.org/10.1080/713668875.

Peccerelli, Fredy. 2012. "Acceptance Speech, Puffin Foundation," 13 May. https://vimeo.com/42718745.

Perreault, Thomas, and Patricia Martin. 2005. "Geographies of Neoliberalism in Latin America." *Environment & Planning A* 37 (2): 191–201. https://doi.org/10.1068/a37394.

Prensa Communitaria. 2016. "Caso CREOMPAZ: militares acusados de desaparición forzada y delitos contra los deberes de humanidad." 11 January. https://comunitariapress.wordpress.com/2016/01/11/caso-creompaz-militares-acusados-de-desaparicion-forzada-y-delitos-contra-los-deberes-de-humanidad/.

Prince, Mischa, dir. 2010. *Still Alive*. Producciones Xocomil.

Recuperación de la Memoria Histórica – Informe Proyecto Interdiocesano
(REMHI). 1998. *Guatemala: Nunca Más*. Guatemala: Oficina de Derechos
Humanos del Arzobispado de Guatemala (ODHAG).

Rodríguez, James. 2010. "Exhumaciones en La Verbena: Llegó la hora, con esta
evidencia, de buscar justicia." MiMundo.org, 23 September. http://www.
mimundo-fotorreportajes.org/2010/09/exhumaciones-en-la-verbena-llego-
la.html.

– 2013a. "Genocide Trial Begins in Guatemala." MiMundo.org, 19 March.
http://blog.mimundo.org/2013/03/2013-03-19-genocide-trial-begins-in-
guatemala/.

– 2013b. "Wartime Victims Exhumed from Former Military Base Return to
Pambach." MiMundo.org, 2 December. http://blog.mimundo.org/
2013/12/2013-11-wartime-victims-exhumed-from-former-military-base-
return-to-pambach.

– 2014a. "El Entierro de Manuel, una Espera de Treinta y Dos Años: Pacux,
Rabinal, Baja Verapaz, Guatemala (14 y 15 de mayo, 2014)." MiMundo.org,
5 June. http://www.mimundo-fotorreportajes.org/2014/06/el-entierro-de-
manuel-una-espera-de.html.

– 2014b. "A Funeral for Manuel: 32 Years after the Guatemalan Massacre That
Killed Him." *Vice News*, 11 June. https://www.vice.com/en_ca/read/a-
funeral-for-manuel-32-years-after-the-guatemalan-massacre-that-killed-
him.

Russell, Grahame. 2014. "Chixoy Dam Reparations to Be Paid Finally, …
Maybe!" *Rights Action*, 17 October. http://questoes.blogs.com/news_
about_mining/2014/10/chixoy-dam-reparations-to-be-paid-finally-maybe.
html.

Schirmer, Jennifer. 1998. *The Guatemalan Military Project: A Violence Called
Democracy*. Philadelphia: University of Pennsylvania Press.

Smith, Emilie. 2012. "Can These Bones Live?" *Sojourners Magazine*, June.
https://sojo.net/magazine/2012/06/can-these-bones-live.

Snow, Clyde C., Fredy A. Peccerelli, José S. Susanávar, Alan G. Robinson, and
Jose Maria Najera Ochoa. 2008. "Hidden in Plain Sight: X.X. Burials and the
Desaparecidos in the Department of Guatemala, 1977–1986." In *Statistical
Methods for Human Rights*, edited by Jana Asher, David Banks, and Fritz J.
Scheuren, 89–116. New York: Springer. https://doi.org/10.1007/978-0-387-
72837-7_5.

Statistics Canada. 2015. "Homicide Offences, Number and Rate, by Province
and Territory (Homicide Rate)." http://www.statcan.gc.ca/tables-
tableaux/sum-som/l01/cst01/legal12b-eng.htm.

teleSUR English. 2016. "Expert: Guatemala Arrest of 14 War Criminals a Major Step," 9 January. https://www.telesurtv.net/english/news/Expert-Guatemala-Arrest-of-14-War-Criminals-a-Major-Step-20160107-0038.html.

Tyner, James. 2012. *Genocide and the Geographical Imagination: Life and Death in Germany, China, and Cambodia.* New York: Rowman & Littlefield.

United Nations Office on Drugs and Crime. 2013a. *Global Study on Homicide: Executive Summary.* Vienna: UNODC. http://www.unodc.org/documents/gsh/pdfs/GLOBAL_HOMICIDE_Report_ExSum.pdf.

– 2013b. *UNODC Homicide Statistics 2013: Homicide Counts and Rates in the Most Populous City, Time Series 2015–2012.* Vienna: UNODC. https://www.unodc.org/gsh/en/data.html.

Weld, Kirsten. 2014. *Paper Cadavers: The Archives of Dictatorship in Guatemala.* Durham, NC: Duke University Press. https://doi.org/10.1215/9780822376583.

3 Transnational and Local Solidarities in the Struggle for Justice: Choc versus Padilla

KALOWATIE DEONANDAN AND
REBECCA TATHAM

Introduction

> I'm going to speak from the bottom of my heart, as an Indigenous woman.
> The one and only thing I want is justice.
>
> (Choc cited in Rights Action 2016a)

This is the plea made by Angélica Choc in the Court of Appeals in Puerto Barrios, Guatemala, on 20 January 2016. She was the main plaintiff in the criminal trial of Mynor Padilla, former head of security at the Fénix nickel mine. Padilla was accused of seriously injuring seven campesinos[1] who were protesting against the mine; shooting a young bystander, Germán Chub Choc (who was paralyzed as a result); and killing Adolfo Ich Chamán, a community leader and anti-mining activist who was also the husband of Angélica Choc. At the time of the incident, the mine was owned by Canada's Hudbay Minerals through its Guatemalan subsidiary, Compañía Guatemalteca de Níquel (CGN; Guatemalan Nickel Company). The trial, which began on 8 April 2015, received scant national and international media attention, despite its significance in terms of justice for Indigenous peoples in Guatemala, including Indigenous women, and the message the alleged crime conveyed about the behaviour of transnational mining companies and their employees. While the unfolding of the case exposed the challenges to Indigenous justice struggles in Guatemala, it also reveals the importance of networks of solidarity for such struggles.

This chapter addresses the prevailing silence surrounding this trial by examining the importance of solidarity for the main plaintiff. It argues that the solidarity networks that have supported Angélica Choc, global and local, have been critical in empowering her in her fight to seek justice for her husband's assassination. At the same time, it draws attention to the limitations of this solidarity in Guatemala, where racism against the Indigenous population is pervasive and entrenched, where elite dominance substitutes for the rule of law, and where corruption and impunity prevails. In making these claims, our analysis draws heavily on the comments and views of Angélica Choc herself. One author attended six courtroom hearings in August and September 2015 and conducted an in-depth interview with Choc herself, as well as one with two legal assistants for the prosecution. A follow-up interview was also done with Choc in February 2017, just weeks before the verdict was due. It is these interviews and the observations of the courtroom proceedings that gave the authors insights into the role that solidarity networks played in Choc's search for justice.

Background

Angélica Choc's struggles are rooted in events that occurred almost sixty years ago. In 1950 the Canadian mining giant INCO arrived in El Estor, on the shores of Lake Izabal in eastern Guatemala. The region is inhabited predominantly by Q'eqchi'-Maya communities. At the time, INCO was a global leader in nickel mining but was beginning to face competition from others such as Sherritt Gordon Mines and Falconbridge Nickel. The company sought to expand its holdings to counter this industry threat. As Bradbury wrote, INCO's expansion into Guatemala was "part of a long-term strategy to retain a dominant position within the global space economy of the nickel industry" (Bradbury 1985, 132).

In partnership with the Hanna Mining Company of Cleveland, Ohio, INCO established the Guatemalan subsidiary Exmibal and began extraction in the region in 1966 (Bradbury 1985, 138). This was the country's first mining operation, and, as with all subsequent ones, it was accompanied by violence, including murder. Before elaborating on the conflict surrounding the mine, it should be noted that the operation changed ownership on several occasions, from INCO to Skye Resources (when the mine became known as the Fénix Project and operated under its new owner's Guatemalan subsidiary, CGN), to Hudbay, and finally to Russia's Solway.[2]

From the outset, Indigenous peoples were opposed to the mine for fear of losing their lands and livelihoods. Guatemalan intellectuals, in particular law professors at the Universidad de San Carlos de Guatemala (University of San Carlos), were concerned that the very generous terms on which the mining concession was granted were detrimental to the interests of Guatemalans, and to the state (Driever 1985; Bradbury 1985), and that outside powers (in this case, Canada) had an unacceptable role in drafting the mining law that allowed these generous terms of investment. Consequently, they began investigating the process that gave rise to the new code (McFarlane 1989, 1–245; Imai, Maheandiran, and Crystal 2014, 1–34) and soon after two of them, Julio Camey Herrera and Adolfo Mijangos López, were assassinated by the military (cited in Ball, Kobrak, and Spirer 1999, 18). Their murders signalled the onset of decades of mining-related violence in the region, which persists still, as the death of Adolfo Ich Chamán reveals.

The local Q'eqchi'-Maya population living in the community of Las Nubes, near the mine site, are regarded as "illegal occupiers" (Hudbay 2009) by the company, which claims to have "legal rights to ... these lands ... based on the concessions it inherited from its predecessors EXIMBAL, INCO and Skye Resources" (Behrens 2009). This position is countered by many of the Q'eqchi'-Maya people in this region who assert ancestral claims to the territory and reject mining development, which they see as compromising the sustainability of their lands and their livelihoods (Nolin and Stephens 2010).

The events that culminated in the current trial began on 27 September 2009, when protests erupted as a result of the "intrusion of Fénix security personnel into Mayan Q'eqchi' communities," and the latter's "fears of renewed forced and violent evictions" (Imai, Maheandiran, and Crystal 2014, 13). As the protest moved towards the town of El Estor, it garnered support from other communities around the Fénix mine, such as La Unión, where Adolfo Ich Chamán lived. Sometime in the early afternoon, violence erupted in three distinct areas, culminating in three different crimes: the shooting murder of Adolfo Ich Chamán, the shooting and paralyzing of Germán Chub Choc, and the wounding of the seven campesinos. While the events that led up to these violent incidents are disputed, with the company presenting one version (see Hudbay 2009) and the community another, many key eyewitnesses told a consistent story. From the outset they identified Mynor Padilla, a retired lieutenant colonel of the Guatemalan Army and the security chief at the mine, as the shooter. Despite an arrest order for him, Padilla fled into hiding and remained a fugitive from justice[3] until he was captured in September 2012.

As the wife of the murder victim, Adolfo Ich Chamán, Angélica Choc was the principal plaintiff in this case. She is a poor Mayan woman who has limited resources to mount a legal fight against Padilla, his high-profile lawyers, and the company that backs him. Yet she persists in this fight in the face of daunting odds, including Guatemala's legendary culture of impunity and its historic and ingrained racism against its Indigenous peoples. The more immediate danger that Choc confronts are the constant threats to her life. Just a few months before the trial ended, for example, armed men fired shots at her home while she and her two youngest children were sleeping inside. Though no one was injured, this incident forced her to relocate to another city with her family as she awaited the verdict.

Solidarity, both transnational (cosmopolitan) and local (communitarian), has been critical to Choc, as she herself attests. At the same time, however, embedded within these solidarities are tensions and contradictions that will, in all likelihood, prevent her from receiving justice.

Theories of Solidarity and Justice

In light of Guatemala's reputation for corruption, elite dominance, violence, racism, and gender inequality (especially against Indigenous women), how is it possible for a woman like Angélica Choc to get her case before the courts? The odds against her are staggering. She is Indigenous, poor, and a victim of violence that is linked to a multinational mining giant that has the backing of the state and the military. Against seemingly insurmountable obstacles, she perseveres in a judicial system plagued with impunity and racism. In so doing she confirms the argument of David Carey Jr that the Guatemalan legal system is "both an extension of the government's coercive and administrative apparatus and, ironically, one the few escape valves available for popular dissent" (2013, 15). However, access to the legal system is particularly onerous for Indigenous peoples in Guatemala, and even more so for Indigenous women. What accounts for Angélica Choc's ability to pursue this route to justice? Her personal strength, stamina, and commitment notwithstanding, the contention here is that her empowerment is rooted largely in the solidarity network that supports her, from the global to the local.

Sally J. Scholz introduces the concept of "political solidarity," through which she extends the sphere of the rights struggles beyond national borders to generate a more cosmopolitan understanding of justice than is allowed for by scholars such as John Rawls (1999, 78–81), who argues that the requirements for justice apply primarily among members of a single nation state. As Scholz notes, political solidarity unites individuals on the basis of their shared commitment to a political cause: "Political solidarity may be motivated by any number of factors such as feelings of indignation, a commitment to justice, experiences of oppression or injustice, desire to care for others who are suffering … it is the mutual commitment, however, that forms the unity of solidarity, not shared feelings, experiences, identities, or social locations" (2007, 38).

Scholz also contends that solidarity incorporates opposition, more specifically "opposition to oppression or injustice" (2007, 38). Through this claim, she aligns her understanding of political solidarity with that of Kurth Bayertz (1999), who argues that the concept can also include a rejection of the policies and tactics practised by another group in the structure within which the solidarity occurs. Hence, "political solidarity aims to effect social change" (Scholz 2007, 40). While Scholz refers largely to political solidarity in social movements, the concept is also applicable to individual actors or organizations (e.g., CICIG's support of Choc). Thus the involvement of transnational groups with no identity ties to Choc's Mayan community can be explained by their shared goal of assisting her in her fight for justice.

As Scholz recognizes, not all actors in a dynamic of political solidarity will have the same degree and type of commitment: "Some people will work diligently and systematically to chip away at what they perceive to be a system of oppression or injustice. Others will suddenly find themselves united by a passion for a cause or incensed by a current injustice. The unity that forms between actors might be close knit and coherent or quite amorphous and fluid" (2007, 40). Some transnational solidarity groups like Rights Action have been unwavering in their support of Choc. Others, such as AcoGuate, have been more sporadic. While the objectives of these two groups coincided in the Padilla trial (both support Indigenous groups), for AcoGuate it is not the focus of their activism, while for Rights Action the struggle against mining is central. While the goals of the two groups overlap, they do not coincide sufficiently to make for permanent collaboration. CICIG, for its part, has a larger goal of promoting justice in Guatemala; this trial was a small component of its more comprehensive project. Yet all three groups, regardless of national character or level of commitment, exemplify transnational solidarity in the quest for justice.

In keeping with the wider scope delineated by Scholz, Carol C. Gould extends political solidarity to the transnational realm through the idea of "network solidarities" (2007, 161). She argues that in the age of globalization, human interconnectedness has expanded significantly, including across territorial space, and so too have the obligations of care, which include a duty to fight against oppression: "The shared values that characterize these solidarity relationships consist [of] a shared commitment to justice ... [and] to the elimination of suffering." Gould clarifies that solidarity stems not just from a "moral disposition" of caring for fellow human beings, but that it also involves a "social critique and attention to institutional structures, as well as to the opportunities that changes in such structures might afford for improving the lot of others" (158). This is only logical, for "if people are to be helpful to others and act to support them, it is useful for those involved in the solidarity relationships to have some idea of the causes of the oppression or suffering" (158). In this way Gould echoes Scholz's claim that political solidarity is imbued with the notion of opposition and that it seeks to bring about change. The virtues of transnational solidarities are also defended by Joseph Schwartz, though he stresses that to be effective, actors must "have sufficient presence" (2007, 132) on the local scene to apply effective pressure to the target state to bring about change.

While Scholz, Gould, and Schwartz shed light on transnational solidarities, the impact of local solidarities in justice struggles should not be ignored. As Toni Erskine states, "Cosmopolitanism neglects the profound importance of local ties and loyalties, community and culture" (Erskine 2007, 125). Her notion of "embedded solidarity" is an attempt to bridge the gap between the respective claims of cosmopolitan and communitarian solidarity. Drawing on insights from feminist and communitarian scholars, Erskine articulates a vision of community as morally constitutive and not necessarily constrained by geographic boundaries. Her claim is that "fellow moral agents" coexist simultaneously in a multiplicity of realms, or as she labels it, where "circles intersect"; they have overlapping community memberships. As she writes, "Embedded cosmopolitanism locates the standpoint of the moral agent in the multifarious communities to which she belongs. The moral agent is thereby not abstracted from all particularity but remains embedded in any number of different morally constitutive associations" (Erskine 2002, 474). The web of intersecting communities in which one can hold membership simultaneously allows for the concomitant coexistence of solidarities globally and locally. In other words, borrowing from Marilyn Friedman (1989), she suggests community obligations are not just "of place" (as in the communitarian emphasis on territorial communities) but may also be "of choice." Similarly, Larry Blum makes a case for the coexistence of what he labels respectively "in-group "and "out-group" solidarity. He asserts, "Solidarity is akin to community" (Blum 2007, 54). However, solidarity and community are not identical, as "one cannot have a sense of community with someone with whom one does not have an ongoing relationship ... the idea of solidarity allows someone to show solidarity with a group with whom this individual does not have such a relationship" (54) and involves "group identity" and "shared experience." The notion of "of-place" or "in-group" solidarity is critical to Choc, in light of her Q'eqchi'-Maya identity. Her fight for justice is rooted in securing justice for herself and her "in-group." As will be demonstrated later in this chapter, cosmopolitan and communitarian approaches to justice can also be contradictory and mutually undermining. Schwartz (2007, 144) reminds us of this when he notes that justice struggles are often constrained by the reprehensible conduct of more powerful actors (e.g., states and institutions) in the system, which are aided by "powerful domestic interests." As well, he draws attention to the restrictions imposed on communitarian solidarities when he notes that "the family, community or tribe can be the site for patriarchy, exploitation, and oppression" (138).

Transnational Solidarity

Three of the most important sources of transnational solidarity for Angélica Choc are the NGOs Rights Action and AcoGuate, and the UN-affiliated government body CICIG.[4]

Rights Action is an international NGO that supports community struggles in Latin America (Rights Action, n.d.b). Its activities focus largely on promoting environmental justice; combatting the harmful impacts of mega development projects (such as mining, biofuels, and dams); aiding the fight against impunity; and providing emergency assistance to disaster-affected communities. Its support includes accompanying at-risk activists, publicizing rights violations committed abroad by corporations in the latter's home countries, and providing direct financial support for those engaged in justice struggles (ibid.). The last has been crucial for Angélica Choc, and she speaks directly of its importance to her and her struggle:

> Un ejemplo en el caso mío son los apoyos pequeños, apoyos pequeños económicos. No voy a decir que hay muchas organizaciones que tienen

conocimiento del caso que estoy llevando o tal vez saben pero se han limitado al apoyo diría yo porque la única organización que me ha apoyado desde el principio hasta hoy en día es la organización de Derecho en Acción. Este, del coordinador Grahame Russell, gracias a él este trabajo se ha avanzado, gracias al apoyo.

También, gracias a las personas porque él no tiene fábrica de dinero. Pero él tiene que hacer mucho para que las personas tengan conocimiento y conciencia de nosotras y cuánto sufrimos y las violaciones de derechos humanos en defensa de nuestro territorio. Y de esto estoy muy agradecido. Y yo sé que en Canadá, y no sólo en Canadá, hay muchas organizaciones que dan sus pequeños granitos de arena con la organización porque a través de Derecho en Acción llegan conmigo los pequenos apoyos. (Choc 2015)

One example of this in my case are the small amounts of help, small amounts of financial help. I won't say that many organizations know about the case I'm bringing, or maybe they know but they've limited their support, because the only organization that has supported me from the beginning up to the present day is Rights Action. Thanks to this organization, coordinated by Grahame Russell, this work has advanced, thanks to his support.

But I am also thankful to all the people who support him because he does not own a money factory. But he has to do a lot to make sure people know about and are conscious of our struggles, of how much we suffer, and all the human rights violations we face in the defence of our land. And for this I am very grateful. I know that in Canada, and not just in Canada, there are many organizations that give "their little grains of sand" to Rights Action, which reach me as small amounts of support.[5]

In addition to its financial support, Rights Action has been publicizing the trial to a Western audience on its Facebook page and in newsletters posted on its website (Rights Action n.d.a, n.d.b). For example, the organization highlights the racism Choc faces in the courtroom, directly from the judge (Rights Action 2016a) and also the threats and intimidations she faces from unknown forces, suspected to be security personnel affiliated with the mine, as an op-ed article by its director, Grahame Russell (2016) demonstrates.

However, in her statement Choc also identifies the limits of this form of solidarity. As she notes, the organization is not a "money factory." It relies heavily on individual donations to conduct its work. In addition, it is driven primarily by the actions and commitments of one individual, Grahame Russell. As such, its reach is limited, especially when compared to that of powerful actors with an abundance of resources, such as transnational mining companies. Nevertheless, the solidarity it provides has been of great significance in contributing to the empowerment of Choc in this struggle.

Another non-governmental organization that provided solidarity support for Choc in this struggle is AcoGuate, which was founded in 2000. It is an international umbrella organization comprising solidarity groups from ten different countries in North America and Europe, whose mandate is to provide "international accompaniment to individuals or organizations from the Guatemalan social and human rights movement who are (or fear being) threatened or harassed as a result of their work against impunity and towards the construction of a democratic, multi-ethnic and pluricultural society that is based on socioeconomic justice and respect for human rights" (Acoguate 2013, 3).

AcoGuate's solidarity comes in more intangible forms than that of Rights Action. By its presence, it seeks to provide international protection to at-risk activists like Choc; for example, during the trial, shots were fired into her home by unknown assailants while she slept with her children (Russell 2016). By being present in the courtroom with Choc, AcoGuate sent a message that it would be a witness to any harms that may come to her; in this way, it offered some safety. The threats faced by Choc are not unusual in Guatemala, as made evident by Magalí Rey Rosa in chapter 4. A U.S. State Department report asserts that intimidation and harassment are rampant in the country's judicial system and targets everyone from judges to prosecutors to plaintiffs (U.S. Department of State 2012); those targeted are usually challenging the dominant power structure and its beneficiaries. Given the rampant impunity in Guatemala, not even the presence of international observers can guarantee that no harm will not come to Angélica Choc. Still, Choc recognizes the importance of this solidarity for her safety:

> Con respeto al caso mío que estoy llevando, yo estoy muy agradecida con el apoyo internacional como observadores internacionales que me han acompañado. Para mí es muy importante y de esto estoy muy agradecida porque así se dan cuenta verdad cómo está la justicia aquí en Guatemala. (Choc 2015)

> With respect to my criminal case, I am very grateful for the international support, such as the international observers who have accompanied me

[to the trial]. For me, this is very important and I am very grateful [for their presence], because that way they realize what justice really looks like here in Guatemala.

Like Rights Action, AcoGuate confronts limitations in the scope of its actions. It cannot prevent harm to Choc; it is merely a dissuasive force. The solidarity of such groups also means that they become targets of delegitimization (Deonandan 2015, 42), or smear campaigns, or worse, victims of violence. The U.S. State Department, citing data from the Unidad de Protección a Defensoras y Defensores de Derechos Humanos de Guatemala (UDEFEGUA; Guatemalan Human Rights Defenders Protection Unit), noted that in 2012, for example, 13 human rights defenders were murdered, while 291 were attacked, and "many of the attacks [were] related to conflicts over land and the exploitation of natural resources" (U.S. Department of State 2012). For the international NGO Peace Brigades International (PBI), these campaigns are "direct attempts to discredit and weaken the human rights movement and to make human rights defenders, organizations, and communities feel more vulnerable" (PBI 2013).

The third source of transnational solidarity for Choc is the CICIG. This was the body established in 2007, with help from the United Nations, to help realize the aims embodied in the 1996 Peace Accords that ended the civil war. More specifically, it is tasked with "the investigation of crimes committed by members of illegal security forces and clandestine security structures ... that affect the fundamental human rights of the citizens of Guatemala" (CICIG n.d.b), and its "mandate permits it to carry out independent investigations [and] to act as a complementary prosecutor" (UNDPA 2014). Aside from its investigative and prosecutorial roles, CICIG also works to promote public policy changes geared to strengthening Guatemala's judicial institutions. As such, it collaborates with the office of the Ministerio Público (MP; Public Prosecutor's Office) and the Policía Nacional Civil (PNC; National Civilian Police) and provides them with technical support in their investigative roles (CICIG n.d.b). One unique feature of CICIG is that it is a hybrid justice mechanism, as "it has many of the attributes of an international prosecutor, but it operates under Guatemalan law, in the Guatemalan courts, and it follows Guatemalan criminal procedure" (CICIG n.d.a). The organization is directly involved in this case, being a joint plaintiff with Choc, "intervening alongside the Office of the Public Prosecutor in the case against Padilla" (Cuffe 2015). CICIG personnel were present each day in the courtroom with Choc.

The solidarity provided by CICIG's presence is dual. The organization's presence not only offers limited protection from harms, as does AcoGuate; it also legitimizes Choc's struggles in a broader national and international context, focusing on the larger justice implications of this trial taking place in a remote town in a small courtroom. CICIG's presence connects the harms suffered by Choc to the relationship between the military (of which Padilla is a representative), transnational capital (CGN), and the political elite that supports them both. Yet CICIG's presence could not bring about a decision that was just, nor could it guarantee her safety. It works within the system in Guatemala and with national actors who may have limited powers, or who may not be interested in a just outcome. As the Washington Office on Latin America (WOLA) notes, "Because the CICIG works hand in hand with the Guatemalan government, its success or failure depends on the political will of the leadership of its counterpart institutions: the MP, the PNC, the judiciary, and the legislature" (WOLA 2015). As the UNHCR cautioned, "Neither Guatemala nor the international community should fall into the trap of seeing CICIG as 'the' solution to Guatemala's failing criminal justice system" (UNHCR 2009, 12). As the discussion below suggests, there is strong evidence that the judge in this case, Ana Leticia Peña Ayala, was strongly biased against the plaintiff, because of race, or perhaps even out of fear for her own safety should she not render a verdict favourable to the defendant.

Aside from the limitations faced by each individual solidarity network, Scholz draws attention to another potential drawback: the fluidity of commitment of each, and the lack of coordination among them. This absence of coordination and unity can be detrimental, in that there may be a multiplicity of actors, but their beneficial impacts are minimal and do not reflect their numbers (Scholz 2007, 40). The wider power structures of transnationalism also mitigate against justice for Choc. These include the forces of neoliberal globalization (the mining companies) and the socio-economic and political structures that defend them, including those that are located within a nation state. Even though, at one level, this was just a local murder case, it had implications for transnational mining companies operating in the country. As J.G. Frynas notes, successful litigation can be costly for companies, both to their reputations and to their profit margins (2004). Hudbay and its subsidiary CGN had a strong interest in getting a ruling favourable to them. There were suspicions that, to this end, they had enabled

Mynor Padilla to secure top-notch legal representation. The defendant initially had three high-profile lawyers; it is not clear how he was able to afford them. One was Francisco José Palomo Tejeda,[6] who had served as a legal representative for General Ríos Montt during the latter's trial for genocide (see chapter 5 by Helen Mack Chang and chapter 6 by Claudia Paz y Paz). The other was Frank Trujillo, who is now implicated in the larger national "la línea" scandal that brought down former president Otto Pérez Molina and his vice-president, Roxana Baldetti, as described by Candace Johnson in chapter 1. The third lawyer, still representing Padilla, was Carlos Rafael Pellecer, who had worked for CGN and Hudbay, and who had been a witness for the company in the lawsuits in Canada (Rights Action n.d.). Two new lawyers joined his team to replace Palomo Tejeda and Trujillo.

For this case in particular, where the plaintiff had a parallel legal action in Canadian courts, the importance of a company victory became even greater. Hudbay brought its considerable resources to bear to mount a vigorous anti-solidarity campaign to delegitimize Choc. This campaign was explicitly in evidence in the Puerto Barrios courtroom in the person of Hubay advocate John Terry, a Harvard-trained partner in the prominent Toronto-based law firm Torys LLP.[7] According to the company's website, Terry has been recognized as one of the best lawyers in Canada, a "leading lawyer in international arbitration … and aboriginal law" (Torys LLP n.d.). His background and status were emblematic of the power structures aligned against Choc in this case, which bridged the Guatemalan and Canadian elites.

While Terry was not officially representing Hudbay in this case, he was, nevertheless, implicitly doing so as he worked on the company's behalf in a related case. Hudbay had hired him to review the affidavits of some plaintiffs, including Angélica Choc and the witnesses on her behalf, who were involved in the civil suit against the company that was pending in the Ontario Supreme Court in Canada. In fact, Terry openly acknowledged in his testimony in the Choc case that Hudbay had paid all his travel expenses to Guatemala. Moreover, in a casual conversation with one author about the challenges of travel in Guatemala, he volunteered that Hudbay had provided private air transport for him within the country so that he could be at the courthouse to testify. What Terry's comments reveal is that Hudbay was willing to invest significant sums of money to ensure he was heard in the courtroom. The goal of his testimony was twofold: to delegitimize Choc and to safeguard the company's reputation.

Terry sought to delegitimize Choc by introducing information on the latter's Canadian lawsuit in such a way as to cast doubt on her credibility and to undermine her character. Through his responses in the courtroom, he called into question Choc's commitment to justice by implying her struggle was all about payment rather than principle. He took every opportunity to stress the sum of the damages she sought in her Canadian suit, that it was "for no less than $12 million." In a country where most Indigenous peasants live on less than $1 per day, this is a staggering and incomprehensible amount. Mention of this massive sum was no doubt calculated to feed the assumptions of the community that Choc was profiting from the trial.

Terry's testimony aligned perfectly with the defence's strategy of delegitimizing Choc by portraying her as mercenary, using the trial as a means for financial gains. One calculating way of doing this was to refer to the lawsuit in Canada and the financial restitution the defendants, including Choc, were seeking. As Padilla's lawyer often sarcastically noted during the trial, Choc had "12 million reasons" (a blatant reference to the Canadian case) for bringing her suit in Guatemala.[8] For Choc herself, as she revealed in an interview just weeks before the Guatemalan verdict was due, this accusation – that she was out only for personal gain – had been one of the most demoralizing and distressing aspects of the trial:

> Ultimamente yo entendí algo que la defensa está diciendo desde hace tiempo en el proceso: Que yo, en esta búsqueda de justicia, llevo un interés personal. Ese tiene nombre y son "12 millones de razones en inglés" decían ellos. Y yo no entendía al principio y yo me preguntaba "Pero qué son 12 millones de razones en inglés?" Y por fin decía: "¿Será que hablan de la demanda que tengo yo en Canadá?" Y eso si me afectó, me afectó mucho ... Y ahora, en las conclusiones de la defensa, el abogado de la defensa dijo: "Esa Señora le ha venido a mentir, Señora juzgadora, y es un delito, Señora juzgadora! Esa Senora, lo que ha hecho es un negocio y eso son los 12 millones de razones en inglés, o, mejor dicho, 12 millones de dolares!" "¿¿Qué??" dije yo y después yo callada, muda. Me dolió tanto, tanto que yo lloré porque mi esposo no tiene precio. No tiene precio para poder decir 12 millones, ¡No! La vida de mi esposo no tiene precio. (Choc 2017)

Recently, I have come to understand something that the defence has been saying for a while now in the trial: That I, in this search for justice, have a personal interest. This [according to the defence] has a name and it is called "12 million reasons in English," they would say. At first, I didn't understand and I would wonder: "What are 12 million reasons in English?" And finally I said, "Could it be they are talking about the lawsuit that I have in Canada?" And this affected me, it affected a lot ... And now, in the defence's concluding remarks, one of the defence lawyers said, "This woman has come to lie, Your Honour; this is a crime, Your Honour! What this woman has done is to go into business [referring to the trial] and these are the '12 million reasons in English,' or, to state it more clearly, 12 million dollars!" "What?" I said, and then I was silent, mute. This hurt me so much, so much that I cried, because my husband's life has no price. It doesn't have a price, especially not 12 million. No! My husband's life doesn't have a price.

It should be noted that even if Choc's Canadian case had an outcome in her favour, and this was highly uncertain, evidence from other contexts reveal that such cases take five to nine years, and "the biggest beneficiaries are the lawyers who often demand hefty fees from their clients" (Frynas 2004, 381–2).[9]

Community Solidarity

For Angélica Choc, as an Indigenous woman, the solidarity provided by belonging to a community with which she shared a group identity rooted in her indigeneity, was a critical source of her empowerment. She, in turn, interpreted her struggles in the framework of protecting the rights of Indigenous peoples. Speaking of her experiences within the courtroom and in the larger Indigenous rights struggles, she proclaimed,

> Pero jamás, nunca hemos estado escuchados aquí en este país, pero yo aún lo estoy viviendo el racismo en la corte, la jueza siempre está al lado de la gente de dinero. Dice ella que nos trata por igual. Yo no soy ciega para darme cuenta, ni también soy muda para no poder hablar aunque les duele que yo les diga sus verdades pero es la verdad. Claro y no tengo estudio pero mis ancestros me dejaron esta sabiduría que ellos tenían y de eso estoy orgullosa, de mis abuelos que se han ido, que cayeron en defensa del territorio y yo no me voy a callar. (Choc 2015)

We [Indigenous peoples] are never listened to in this country, and I am still experiencing this racism in the courtroom through the judge, who is clearly on the side of people with money. She says she treats us all equally. But I'm not blind and I see what's going on, nor am I mute and unable to speak. And regardless of who I may offend, I tell them the truth, because that is the truth. Of course I never studied, but my ancestors left me their wisdom and I am very proud of my ancestors who have passed on, who fell in defence of our land. I will not be quiet.

In her quest for justice, she was supported by several solidarity groups[10] "of-place"; not all were Indigenous. She spoke, for example, of the support from other mining resistance movements in the country and the strength she derived from it:

Sí, y gracias también pues que va la fuerza que tengo. Muchas de las resistencias me han invitado a sus comunidades o pueblos. He llegado a la Puya para los aniversarios y a San Rafael las Flores ... y tambien a la Mina Marlin porque los compañeros me han llamado ... siempre para compartir las experiencias. (Choc 2015)

Yes, I have received a lot of my strength from them [referring to other anti-mining struggles in Guatemala]. Many of the resistance movements have invited me to their communities or villages. I have gone to La Puya for anniversary celebrations and to San Rafael las Flores ... and also the Marlin mine, because my friends called and asked me to go ... always to share our experiences.

The resistance movements in La Puya and San Rafael las Flores are led by the non-Indigenous groups. The Marlin mine resistance is predominantly Mam-Mayan. These struggles are described in detail by Magalí Rey Rosa in chapter 4. Aside from anti-mining movements, Angélica's struggle is backed by national organizations that provide her with support as part of their solidarity commitments. One of the most important of these is the Comité de Unidad Campesina Peasant Unity Committee (CUC; Peasant Unity Committee). CUC was the first national organization formed by peasants and Indigenous people in Guatemala, in 1978, in response to the violence unleashed by the country's military dictatorships. Over the last four decades, the organization has grown significantly. It now has representation in over two hundred communities in the country (CUC n.d.). Its growth has been not only in size, but also in its mandate; it is now active in support of "farmworkers' rights, community-based development, identity and rights of indigenous peoples, gender equity and organizational strengthening" (Grassroots International n.d.).

As part of this expanded role, CUC is now involved in defending Indigenous rights against incursions by mega-project development, including mining, and one direct way in which its solidarity is expressed is very tangible, that is, through the provision of legal support. Choc's legal representative, Sergio Beltetón,[11] was a lawyer in the employ of CUC (Choc 2015). This act of solidarity by CUC was immensely significant in aiding Choc to continue her struggle, as lack of economic resources was one of the greatest barriers to achieving justice in Guatemala, particularly for women (OACNUDH n.d.; Sieder and Sierra 2010). Embodied in the solidarity offered by CUC is the broader "in-group" solidarity – the solidarity of co-nationals. In 2016, CUC organized a major national rally, "La Marcha por el Agua, la Madre Tierra, el Territorio y la Vida" (March for Water, Mother Earth, Territory and Life), or "La Marcha," as it was popularly called, in which approximately eighty grassroots organizations representing Indigenous peoples, took part (Radio Zapatista 2016). The objective of this campaign, as its label suggests, was to draw attention to the harms that large-scale transnational operations in the extractive sector (as well as in the palm oil industry and in hydroelectricity) bring to the local communities. The protest was also to denounce "la criminalización, persecución y judicialización contra líderes comunitarios y defensores de derechos humanos de los pueblos indígenas" (TeleSUR 2016; the criminalization, persecution, and trials against community leaders and defenders of the human rights of Indigenous peoples).

While the mobilization was not explicitly on behalf of Angélica Choc or the Q'echi' communities of El Estor, the aims embodied in the struggles of the latter two very much part of La Marcha's campaign, and thus lent credibility to Choc's struggles. Our initial interviews with Choc occurred before La Marcha, but she would have seen it as a source of solidarity. Choc's empowerment is rooted in community solidarity, which is linked to her Q'echi'-Maya identity. However, there are also undeniable tensions within her community that diminish her empowerment. As Schwartz notes, "The family, community, or tribe can be the site for patriarchy, exploitation, and oppression" (2007, 138).

Mining development in Guatemala occurs largely on Indigenous territories and within a context marked by overwhelming poverty. For those who can better their existence as a result of jobs at the mine, any challenge to the survival of the industry represents a threat to livelihood, in spite of the damage mining does to the environment. Community members who are employed by the Fénix project view Angélica's resistance and quest for justice as a threat to their economic survival, and this leads to tensions between her and them, and to divisions in the community. The source of solidarity for Angélica at the local level is riddled with tensions and divisions. In this context, Frynas's comments on Nigeria apply also to Guatemala: "In extreme cases, the costs of potential future litigation could discourage firms from continuing with a harmful activity in a community and prompt a firm's withdrawal. But the local community may often want the company to stay rather than withdraw, given its reliance on an offending firm in the absence of alternative jobs and income opportunities" (2004, 380).

In Angélica's case, this disunity at the local level is exacerbated by her international profile, which stems from the attention she has received from international NGOs and from her legal case in Canada (Deonandan and Tatham 2016). Many in her community believe she is already receiving massive sums of money, and the authors have witnessed, on more than one occasion, her frustrating attempts to explain to the group that she has not received any funds. She is not the only one who faces such accusations. The eleven alleged rape victims whose case is also before the Canadian courts face similar allegations by their community. These charges, as Choc tells us, seem to be part of an orchestrated strategy of mine supporters to delegitimize resistance members and foster community disunity:

> Pues Lote 8 también estaba fuertemente en contra pero llegó cuando hicieron la demanda las compañeras pero el personal de la empresa a meter mentiras con la comunidad, con los líderes primero diciendo que las mujeres habían recibido una cantidad de dinero que por lo tanto toda la gente vende a desalojar el área. Entonces, ¿qué hicieron ellos? Fueron a las asambleas diciendo que las mujeres que demandaron ya recibieron el dinero y nos vamos a ir y por qué vamos a hacer esto. Entonces, lo que empezaron fue el conflicto en Lote 8. Entonces, esos que escucharon a esas mentiras se fueron hasta trabajar en la mina. (Choc 2015)

> Lote 8 used to be firmly opposed [to the mine], but when the women comrades filed their lawsuit [in Canada], personnel from the mine started spreading lies in the community. First, it was with the leaders [of their community] saying that the women had received a lot of money and therefore all of them should sell and leave the area. So what did they do?

They went to the meetings saying that the women who filed the lawsuits had already received the money and were going to leave. So they started a conflict in Lote 8. So the ones who listened to these lies went to work for the mine.

As Choc tried to broaden the locus of the struggle through the Canadian lawsuit, she was concomitantly restrained by local jealousies that worked to the detriment of local solidarity, jealousies exploited and exacerbated by her transnational opponents. (In chapter 1, Candace Johnson describes the marginalization and exclusion Angélica Choc endures as a result of her struggles.) Local solidarities were also being undermined by witnesses recanting their testimony that Padilla was responsible for the crimes. Both Hudbay and Choc agreed that there were retractions but give different reasons. According to Hudbay, "Three of the four initial 'eye witnesses' reversed their initial claims, acknowledging they did not see anyone shoot Aldofo Ich Chamán and do not know how he was killed. Some now say they were intimidated into making their original claims" (Hudbay n.d.). For Choc, the recanting had more to do with witnesses being bought by the company (Solis 2015; Choc 2015). As she confided, with reference to several witnesses who changed their testimonies,

> Pues si, ella … fue mi testigo también … y dos mujeres más … pero, ellas se retiraron porque la empresa las contactó y ahora testifican a favor de la empresa, a favor del asesino … Les dan víveres mensuales y les dan dinero como ellas mismas lo hablan. (Choc 2015)

> Well yes, she … used to be my witness also … and two more women … but they withdrew [as witnesses] because the mine contacted them and now they are testifying in favour of the mine, in favour of the murderer … they [the mine] give them monthly food bags and money, as they themselves tell.

According to Choc, the company even tried to buy her silence. As she revealed, "Me ofrecieron becas para mis hijos, trabajo para mis hijos, y sí cantidad de dinero. La vida de mi esposo no tiene precio" (Choc 2015; They offered me scholarships for my children, work for my children, and a sum of money. My husband's life does not have a price).

Another significant problem that undermined local solidarity for Choc was endemic racism within the justice system, especially against

Indigenous women, as well documented by the Defensoría de la Mujer Indígena (DMI; Department of Protection of Indigenous Women's Rights) (Sieder and Sierra 2010). Choc has called attention to the behaviour of the presiding judge, Peña Ayala: "I have been discriminated against ever since this trial started. And the people who have been coming to the hearings to support me, other Indigenous people, have been discriminated against too. They claim that these slights were not directed at me. But of course they were directed at me! I am an Indigenous, Maya-Q'echi' woman" (Choc cited in Rights Action 2016a). Isabel Solis, a legal observer for the prosecution, said that initially they believed that the appointment of this judge signalled a chance of fair treatment in spite of the historical evidence of the system's biases: "Al inicio tuvimos mucha esperanza … nos habían dicho que era una jueza muy recta, muy buena, y muy objetiva" (Solis 2015; At the beginning we had a lot of hope … we were told that she was a very strict, good, and unbiased judge).

After witnessing the judge's behaviour, however, Solis came to a different conclusion: "Pero ya con el tiempo vimos que ella [Peña Ayala] no es tan objetiva que se puede decir porque ya varios incidentes y varios elementos nos indican que ella está inclinándose hacia Padilla o la empresa" (Solis 2015; But with time we saw that she was not as impartial as we thought, and various incidents indicate that she is favouring Padilla or the company [Hudbay]). Examples of the alleged discrimination were reported by Rights Action and witnessed by one author of this chapter. One example of Judge Peña Ayala's pro-defendant bias relates to Padilla's accommodation. Normally, he would be housed in a Guatemala City prison. The trial was occurring in Puerto Barrios, for two days at a time. As it was too far to travel to Guatemala City, after each day Padilla was transferred overnight to a holding cell in the nearby Department of Zacapa, which had appropriate facilities. Padilla requested he be allowed to remain in Puerto Barrios, claiming the trip to Zacapa was too tiring for him and his family. Despite the protestations of the plaintiff, the judge granted him this privilege, stating the accused "has rights" (Peña Ayala 2015).

On another occasion, the accused and his lawyers convinced the judge that he "was being harassed by people entering the court" (Rights Action 2016c). According to the video of the incident, Padilla approached a delegation of activists and shook the hands of some. Inside the courtroom, however, he claimed that he was being threatened, though the video showed no evidence of this allegation, and therefore was in need of protection.[12] The judge granted his request, assigning him a security detail. In responding to this situation, Choc complained,

> It makes me so angry that the judge concludes that the man who murdered my husband is the one being threatened. His lawyers come to the trial with armed private security guards all the time; the CGN has had other armed men who have followed us at different times during the trial; there are police everywhere, always escorting [Padilla]; and we have no weapons at all and no security guards at all, ever … and yet she determines that it is [Padilla] who needs police protection! How can I get justice? (Rights Action 2016c)

It is possible that the judge's blatantly pro-defendant bias was the result of political pressures being exerted on her. While no direct evidence was found to indicate that this was so, several factors made it a possibility. The defendant, Mynor Padilla, was affiliated with one of the most powerful, notorious, and feared institutions in the country – the military. He was a retired lieutenant colonel of the Guatemalan Army and would thus have its backing. This support was likely expressed through the Asociación de Veteranos Militares de Guatemala (AVEMILGUA; Association of Military Veterans of Guatemala), a lobby group for ex-military members (Lawyers Committee n.d.) suspected of having strong links to private security forces, such as Delta Elite (Waxenecker 2014, 7), the security contractor for Hudbay minerals. Further, the defendant was in the employ of another powerful entity, the transnational mining company, for which he was head of security. It follows that the company had a vested interest in seeing him exonerated, as a guilty verdict would be a blemish on its reputation.

As well, the disreputable ethics of the Guatemalan legal system made political pressure on the judge a possibility. As noted earlier, Guatemalan courts are infamous for their corruption. As reported by the Bertelsmann Foundation and the U.S. Department of State, "The judicial sector is plagued by widespread corruption and … subject to political influences.… Some judges have reported that they have received 'instructions' on how to rule in specific cases if they wish to remain in their positions … judges are regularly exposed to bribes, intimidation, and violence" (cited in Business Anti-Corruption Portal n.d.). The context in which the judge performed her duties was an additional factor that undermined local solidarity, and the chances of receiving justice, for Angélica Choc.

Conclusion

This analysis has emphasized the roles of different forms of solidarities in aiding Angélica Choc's struggle for justice, and the forces amassed against these networks. Her powerful personal motivations and strength of character are also crucial. Theories of postcolonial feminism remind us of the agency of women in the Global South and reject any sweeping portrayal of them as victims (Mohanty 2003). While Choc had been victimized, she did not see herself as a passive victim, and she spoke eloquently of her own agency and the motivations behind her actions.

Como me ha costado a mí la muerte de mi esposo. Ver a mis hijos llorar, ver a mis hijos cuando ellos cayeron al vicio por la gran tristeza de la desesperación de no tener un padre. Eso me duele como madre, como esposa, como mujer. Siento que ese derecho pues, me lo violaron. Le violaron al derecho de mis hijos de tener un padre ... pero, la lucha de él cayó en mis manos cuando él cayó, cuando lo mataron. Y esa lucha yo la sigo con la fuerza de él que me dejó. Yo voy a seguir luchando no importa cómo. (Choc 2015)

How my husband's death hurt me! Seeing my children cry, seeing them succumb to vices due to the sadness and despair they feel at the loss of their father. This hurts me as a mother, wife, and woman. I feel that they have violated this right of mine. They have also violated the right of my children to have a father ... but, his [Adolfo's] struggle fell into my hands the day he fell, the day they murdered him. I will continue this struggle with the strength that he left me. I will continue fighting no matter what.

In spite of the biases she saw in the system, perceived through the microcosm of the courtroom, she did not cower. Like the women to whom Carey refers, she, too, was using the courts to demand her rights:

Y por eso, yo he puesto mis demandas como mujer, yo sé que tengo derecho y como mujer indígena yo sé que aunque se me han discriminado, me han humillado por demandar, por buscar justicia. Pero aún yo tengo esta energía al cien por ciento para poder seguir adelante, aunque yo me he enfermado pero la muerte de él yo no lo voy a dejar así. Voy a seguir luchando. (Choc 2015)

And this is why I have brought my demands [before the courts] as a woman. I know that I have rights and, as an Indigenous woman, I know that, even though they [the courts] have discriminated against me and humiliated me for seeking and demanding justice. But I still have 100 per cent of my energy to move forward, although I myself have fallen ill, but I will not leave his [Adolfo's] death like this. I will continue fighting.

Angélica Choc brought great strength and perseverance to this battle; but her struggle for justice could not have endured without the intricate network of local and transnational solidarities that support her.

Postscript

Acquitted! This was the verdict handed down by the judge, Ana Leticia Peña Ayala, on 6 April 2017, in the Mynor Padilla murder and aggravated assault charges. Though disheartening, the ruling was unsurprising, as it reaffirmed for Angélica Choc and the other plaintiffs in this case the oft-repeated claims regarding rampant corruption and impunity in Guatemala's judicial system, particularly when it comes to justice for the country's Indigenous groups. Compounding the plaintiffs' disillusionment were the series of post-verdict actions by the judge. She not only apologized to Padilla (she actually asked his forgiveness – "le pidió perdón") for the time he spent in detention as a result of the charges (Solano 2017), but more shockingly, she ordered criminal charges against ten of the people involved in the case against him including Choc, the lead plaintiff, her children and sister, the witnesses who testified on her behalf, other alleged victims in the case, and the prosecuting lawyers from the Attorney General's office (Rights Action 2017a). Giving false testimony, obstructing justice, forging of public documents, and document tampering were some of the charges that Judge Peña Ayala filed against the plaintiffs and witnesses (Rights Action 2017b).

While this stunning turn of events was a blow to the prosecutor's case, it did not go unanswered. The solidarity network, both local and global, that has empowered Choc in her justice struggles immediately countered with their own legal challenge. The office of the country's Attorney General and the UN-backed anti-corruption agency, CICIG, not only filed an appeal against the Padilla verdict, but responded to Peña Ayala's charges by filing counter-charges of their own on 1 June 2017 (Solano 2017). Furthermore, in September 2017, the Court of Appeals in Guatemala City, in a surprising move, ruled that Mynor Padilla would have to face a new murder trial in 2018.

These actions have given hope to the plaintiffs in the case that deep cracks are occurring in the foundations of Guatemala's system of impunity that could one day lead to the system's collapse. Admittedly, however, that time still seems to be in the very distant future as Ivan Velasquez, head of the CICIG noted: "This wounded giant, this criminal superstructure, is difficult to eradicate ... It is something deeply embedded, a strategy designed to keep state institutions working for the benefit of a selected few" (Telesur 2017).

NOTES

1 The injured were: Haroldo Cucul, Santos Caal Beb, Alejandro Acte, Ricardo Acte, Samuel Coc, Alfredo Tzi and Luciano Ical (Small and Jefferson 2014).

2 Aside from Solway, all owners were Canadian. INCO sold the mine to Skye in 2004. The latter sold it to Hudbay in 2008, and then it was sold to Solway in 2011.

3 However, he remained on the company's payroll for at least another year (Small and Jefferson 2014).

4 Other transnational groups active on Choc's behalf include MiningWatch Canada, the Washington-based Center for International Environmental Law (CIEL) and a host of activist NGOs that publicize the trial and help focus international attention on Choc's cause.

5 Translations from the Spanish are by Rebecca Tatham and Stephen Henighan.

6 In 2015 Palomo Tejeda was shot and killed, allegedly by a hit squad. There is no evidence that his death is linked to this case.

7 This is the family law firm of the current mayor of Toronto and former leader of the Ontario Progressive Conservative (PC) Party, John Tory. Tory was the PC Party leader from 2004 to 2009 and was a partner in the firm from 1986 to 1995.

8 The implication is that if Choc wins in the Guatemalan case, this will strengthen her chances in the Canadian lawsuit.

9 In this case, Klippensteins Barristers and Solicitors law firm, which is representing Choc, is working pro-bono. It is not clear what their fees will be in the case of a ruling in her favour.

10 Other national organizations which support her struggle include UDEFEGUA, the Oficina de Derechos Humanos del Arzobispado de Guatemala (ODHAG; Human Rights Office of the Archbishop of Guatemala City), and the Centro de Acción Legal Ambiental y Social de Guatemala (CALAS; Guatemalan Environmental and Social Legal Action Centre),

among others. The groups all signed a petition supporting Choc and the other women in their case in Canada against Hudbay: "We express our recognition to the group of eleven women who survived the sexual assault, to Angélica Choc, widow of Adolfo Ich, and to Germán Chub, for having the courage to demand justice from a powerful transnational enterprise, despite the enormous power and resource inequalities between them, and despite the repressive policies of the GNC, in conspiracy with the State of Guatemala" (cited in Nobel Women's Initiative 2013).

11 As of late 2015, Sergio Beltetón was no longer involved with this case and another lawyer substituted for him.

12 The Judge subsequently made a ruling barring international observers from the courtroom. It seems doubtful that this was due to her concerns for Padilla's safety but rather more to her concerns that her decisions may be subject to criticisms by international observers.

REFERENCES

AcoGuate (International Accompaniment Project in Guatemala). 2013. "The Genocide Case: Observation Report." https://acoguate.files.wordpress.com/2014/09/informecasogenocidioingles.pdf.

Ball, Patrick, Paul Kobrak, and Herbert Spirer. 1999. "State Violence in Guatemala, 1960–1996: Quantitative Reflection." American Association for the Advancement of Science. https://www.aaas.org/sites/default/files/migrate/uploads/Guatemala_en.pdf.

Bayertz, Kurt. 1999. "Four Uses of 'Solidarity.'" In Solidarity, edited by Kurth Bayertz, 3–28. Dordrecht, Netherlands: Kluwer. https://doi.org/10.1007/978-94-015-9245-1_1.

Behrens, Susan Fitzpatrick. 2009. "Nickel for Your Life: Q'eqchi' Communities Take On Mining Companies in Guatemala." North American Congress on Latin America. https://nacla.org/node/6177.

Blum, Lawrence. 2007. "Three Kinds of Race-Related Solidarity." Journal of Social Philosophy 38 (1): 53–72. https://doi.org/10.1111/j.1467-9833.2007.00366.x.

Bradbury, J.H. 1985. "International Movements and Crises in Resource Oriented Companies: The Case of Inco in the Nickel Sector." Economic Geography 61 (2): 129–38. https://doi.org/10.2307/143868.

Business Anti-Corruption Portal. n.d. "Guatemala: Judicial System." http://www.business-anti-corruption.com/country-profiles/the-americas/guatemala. Accessed 22 May 2016.

Carey Jr, David. 2013. *I Ask for Justice*. Austin: University of Texas Press.

Choc, Angélica. 2015. Interviewed by Rebecca Tatham, El Estor, Guatemala, 6 August.

– 2017. Interviewed by Rebecca Tatham, undisclosed location (at interviewee's request), 24 February.

Comité de Unidad Campesina (CUC; Peasant Unity Committee). n.d. Homepage. http://www.cuc.org.gt/web25/index.php?option=com_conten t&view=category&layout=blog&id=10&Itemid=101.

Cuffe, Sandra. 2015. "Nickel Mine, Lead Bullets: Maya Q'eqchi' Seek Justice in Guatemala and Canada." *Intercontinental Cry*, 21 May. https:// intercontinentalcry.org/nickel-mine-lead-bullets/.

Deonandan, Kalowatie. 2015. "Evaluating the Effectiveness of the Anti-Mining Movement in Guatemala: The Role of Political Opportunities and Message Framing." *Canadian Journal of Latin American and Caribbean Studies* 40 (1): 27–47.

Deonandan, Kalowatie, and Rebecca Tatham. 2016. "The Unexplored Dimensions of Resistance to Extractivism in Latin America: The Role of Women and NGOs." In *Mining in Latin America: Critical Approaches to the New Extraction*, edited by Kalowatie Deonandan and Michael Dougherty, 275–85. Abingdon: Routledge.

Driever, Steven. 1985. "The Role of Lateritic Nickel Mining in Latin American Countries with Special Reference to Exmibal in Guatemala." *GeoJournal* 11 (1): 29–42. https://doi.org/10.1007/BF00572937.

Erskine, Toni. 2002. "'Citizen of Nowhere' or 'The Point Where Circles Intersect'? Impartialist and Embedded Cosmopolitanisms." *Review of International Studies* 28 (3): 457–78. https://doi.org/10.1017/S0260210502004576.

– 2007. "Qualifying Cosmopolitanism? Solidarity, Criticism, and Michael Walzer's 'View from the Cave.'" *International Politics* 44 (1): 125–49. https:// doi.org/10.1057/palgrave.ip.8800162.

Friedman, Marilyn. 1989. "Feminism and Modern Friendship: Dislocating the Community." *Ethics* 99 (2): 275–90. https://doi.org/10.1086/293066.

Frynas, Jędrzej George. 2004. "Social and Environmental Litigation against Transnational Firms in Africa." *Journal of Modern African Studies* 42 (3): 363–88.

Gould, Carol C. 2007. "Transnational Solidarities." *Journal of Social Philosophy* 38 (1): 148–64. https://doi.org/10.1111/j.1467-9833.2007.00371.x.

HudBay Minerals. 2009. "HudBay Minerals Provides Update on CGN." News release. http://www.hudbayminerals.com/default. aspx?SectionId=5cc5ecae-6c48-4521-a1ad-480e593e4835&LanguageId=1& PressReleaseId=ee8e6a7d-64d9-4d31-8422-e294df31ca0d.

– n.d. "Guatemala Lawsuits: Factors to Consider." http://www. hudbayminerals.com/English/Responsibility/The-Facts-CGN-and-Hudbay-in-Guatemala/Guatemala-Lawsuits-Factors-to-Consider/.

Imai, Shin, Bernadette Maheandiran, and Valerie Crystal. 2014. "Accountability across Borders: Mining in Guatemala and the Canadian Justice System." Osgoode Legal Studies Research Paper Series. Paper 73. http://digitalcommons.osgoode.yorku.ca/cgi/viewcontent. cgi?article=1028&context=clpe Accessed 15 May 2016.

International Commission against Impunity in Guatemala (CICIG). n.d.a. "About CICIG." http://cicig.org/index.php?page=about.</eref>

– n.d.b. "Mandate." http://www.cicig.org/index.php?page=mandate.

Lawyers Committee for Human Rights. n.d. "A Test for Justice in Guatemala: The Myrna Mack Murder Trial." Lawyers Committee for Human Rights. https://www.humanrightsfirst.org/wp-content/uploads/pdf/test_of_justice.pdf.

McFarlane, Peter. 1989. *Northern Shadows: Canadians and Central America.* Toronto: Between the Lines.

Mohanty, Chandra Talpade. 2003. "'Under Western Eyes' Revisited: Feminist Solidarity through Anticapitalist Struggles." *Signs* 28 (2): 499–535. https://doi.org/10.1086/342914.

Nobel Women's Initiative. 2013. "Hudbay Minerals Faces Canadian Trial for Human Rights Abuses in Guatemala," 1 August. http://nobelwomensinitiative.org/hudbay-minerals-faces-canadian-trial-for-human-rights-abuses-in-guatemala/.

Nolin, Catherine, and Jacqui Stephens. 2010. "'We have to protect the investors': 'Development' & Canadian Mining Companies in Guatemala." *Journal of Rural and Community Development* 5 (3): 37–70.

Oficina del Alto Comisionado de las Naciones Unidas para los Derechos Humanos en Guatemala (OACNUDH). n.d. "Violencia contra las mujeres indigenas en Guatemala." http://www.ohchr.org.gt/documentos/publicaciones/Estudio%20sobre%20violencia%20dom%C3%A9stica%20hacia%20mujeres%20ind%C3%ADgenas.pdf.

Peace Brigades International (PBI). 2013. "Increased Attacks against Human Rights Defenders in Guatemala," 31 May. https://peacebrigades.org.uk/content/increased-attacks-against-human-rights-defenders-guatemala.

Peña Ayala, Ana Leticia. 2015. Proceedings in the Courtroom in the Trial of Mynor Padilla. 4 August. Audio in authors' collection.

Radio Zapatista. 2016. "Crónica de la Marcha por el Agua, la Madre Tierra, el Territorio y la Vida en Guatemala." 21 April. http://radiozapatista.org/?p=16518&lang=en.

Rawls, John. 1999. *The Law of Peoples*. Cambridge, MA: Harvard University Press.

Rights Action. n.d.a. Facebook page. www.facebook.com/RightsAction.org.

– n.d.b. Homepage. rightsaction.org.

– 2016a. "Caso Mynor Padilla Trial: Discriminación Racial // Racial Discrimination." February. https://www.youtube.com/watch?v=Q_nLF-bke1k&feature=youtu.be.

– 2016b. "Mynor Padilla, Hudbay Mineral's Former Head of Security, Is Provided with Police 'Security Detail' for Safety." 18 May. http://unbc. fieldschools.ca/wp-content/uploads/2016/05/CDN-mining-emergency-delegation-to-guatemala-Alert-4.pdf.

– 2017a. "Mining Repression and Impunity in Guatemala: Killers Go Free, Victims Are Accused." 13 April. http://rightsaction.org/newsletterapril17-miningrepressionandimpunity/?utm_source=Rights+Action+Newslette r&utm_campaign=82ca676e5b-EMAIL_CAMPAIGN_2017_04_13&utm_medium=email&utm_term=0_66874e83b6-82ca676e5b-328632993.

– 2017b. "Mining Repression and Impunity in Guatemala: Killers Go Free, Victims Are Accused." 17 April. http://rightsaction.org/newsletterapril17-miningrepressionandimpunity/?utm_source=Rights+Action+Newslette r&utm_campaign=82ca676e5b-EMAIL_CAMPAIGN_2017_04_13&utm_medium=email&utm_term=0_66874e83b6-82ca676e5b-328632993.

Russell, Grahame. 2016. "7 Years Fighting against Canadian Mining Impunity in Guatemala." 19 October. https://www.telesurtv.net/english/opinion/7-Years-Fighting-Against-Canadian-Mining-Impunity-in-Guatemala-20161018-0021.html.

Schnoor, Steve. 2016. "Accused Murderer Mynor Padilla Entering the Courthouse. Puerto Barrios, Guatemala," 17 May. https://www.youtube.com/watch?v=rlIt5geTIJc Accessed.

Scholz, Sally J. 2007. "Political Solidarity and Violent Resistance." *Journal of Social Philosophy* 38 (1): 38–52. https://doi.org/10.1111/j.1467-9833.2007.00365.x.

Schwartz, Joseph M. 2007. "From Domestic to Global Solidarity: The Dialectic of the Particular and Universal in the Building of Social Solidarity." *Journal of Social Philosophy* 38 (1): 131–47. https://doi.org/10.1111/j.1467-9833.2007.00370.x.

Sieder, R., and M.T. Sierra. 2010. "Indigenous Women's Access to Justice in Latin America." Chr. Michelsen Institute (CMI) Working Paper. https://www.cmi. no/publications/3880-indigenous-womens-access-to-justice-in-latin.

Small, Rachel, and Joanne Jefferson. 2014. "Communiqué: Call for Solidarity as Criminal Trial against Mynor Padilla, Former Head of Security for

Hudbay Minerals/CGN, Begins in April." 28 February. Under-Mining Guate. https://rachelblumesmall.wordpress.com/2014/02/28/communique-call-for-solidarity-as-criminal-trial-against-mynor-padilla-former-head-of-security-for-hudbay-mineralscgn-begins-in-april/.

Solano, Luis. 2017. "MP y CICIG accionan contra libertad de militar vinculado a minera en El Estor." 7 June. https://cmiguate.org/mp-y-cicig-accionan-contra-libertad-de-militar-vinculado-a-minera-en-el-estor/?utm_sour ce=Rights+Action+Newsletter&utm_campaign=d6b2910c07-EMAIL_CAMPAIGN_2017_09_20&utm_medium=email&utm_term=0_66874e83b6-d6b2910c07-328632993.

Solis, Isabel. 2015. (Legal Observer for the Prosecution in the Mynor Padilla Case.) Interviewed by Rebecca Tatham, Guatemala City, 4 September.

TeleSUR. 2016. "Marcha por el Agua llega a su fin este viernes en Guatemala," 22 April. https://www.telesurtv.net/news/Marcha-por-el-Agua-llega-a-su-fin-este-viernes-en-Guatemala-20160422-0023.html.

– 2017. "Canada Mining Head Cleared of Guatemala Campesino Killings." 6 April. https://www.telesurtv.net/english/news/Canada-Mining-Head-Cleared-of- Guatemala-Campesino-Killings-20170406-0032.html.

Torys LLP. n.d. "John A. Terry." https://www.torys.com/people/terry-john-a.

United Nations Department of Political Affairs (UNDPA). 2014. "CICIG (International Commission against Impunity in Guatemala)." http://www.un.org/undpa/en/americas/cicig.

United Nations Human Rights Council (UNHCR). 2009. "Promotion and Protection of All Human Rights, Civil, Political, Economic, Social and Cultural Rights, Including the Right to Development: Report of the Special Rapporteur on Extrajudicial, Summary or Arbitrary Executions, Philip Alston, Addendum: Follow-up to Country Recommendations – Guatemala," 4 May. A/HRC/11/2/Add.7. http://www.refworld.org/docid/4a0932270.html.

United States Department of State. 2012. "Country Reports on Human Rights Practices for 2012." https://www.state.gov/j/drl/rls/hrrpt/2012humanrightsreport/index.htm#wrapper.

Washington Office on Latin America (WOLA). 2015. "WOLA Report on the International Commission against Impunity in Guatemala (CICIG). https://www.wola.org/publications/WOLA_report_international_commission_against_impunity_guatemala.

Waxenecker, Harald. 2014. "Poderes Fácticos y la Disputa por los Recursos Estratégicos: Redes, Poder y Violencia." May. https://mx.boell.org/sites/default/files/redes_poder_y_violencia_mayo_2014.pdf.

PART TWO

Justice in Practice

4 A Diary of Canadian Mining in Guatemala, 2004–2013

MAGALÍ REY ROSA

2004

January: Land distribution is one of Guatemala's most sensitive problems. Now that the mining companies have set their eyes on our country, the problem is going to grow more acute. According to Guatemalan law, land may be owned privately, but the subsoil belongs to the state. This means that those who are observing democratic norms for acquiring a parcel of land may be obliged to give it up in order to benefit the transnational companies that plan to exploit our mineral resources. Mining activity will attract investment, it will create some jobs and economic benefits for some people. But it is a process that destroys entire mountains and permanently contaminates the vast quantities of water that it uses, and the soil that peasants and Indigenous people need for their agricultural activity.

More than 200 requests for mining licences have been made in 15 departments of Guatemala. For our peasant and Indigenous brothers throughout the country, this is a warning sign. In Izabal, for example, the mining companies are handing out pamphlets to persuade small landowners to let them use their land for mining. No doubt the companies believe that in this way they are informing the people of their intentions. What they don't say is that after they have been mined, these lands will be useless. We can only hope that our new government won't try to convert Guatemala into a mining country, as this would make them into accomplices in the surrender, poisoning, and destruction of our land.

February: The well-being of Guatemala's population, much of which suffers from serious deficiencies in health care, education, and opportunities for a dignified life depends in large part on our natural and cultural heritage. Someone who is desperate may be tempted by a job as a dynamiter in an open-pit mine – dangerous, poorly paid, short-term work. It is unjust, as is now happening in Izabal, and in many parts of the highlands, to offer those who have so little, money for their land under the threat that if they don't sell, they will be expropriated because the subsoil, which has cost the people of Guatemala such suffering, has been granted in concessions to transnational extraction companies. It is perverse to claim as beneficial activities that destroy the jungles, forests, and mountains, contaminate the water and the soil, and make people ill. Crimes against the environment are reported, but in the majority of cases the investigations are inadequate; prosecution and conviction almost never occur. Behind the crimes and violations of the law are secret private security apparatuses. The Guatemalan state has been unable to guarantee our most basic rights.

National sovereignty also becomes an issue. Other countries' interest in taking advantage of our natural resources has always been strong, and it will grow stronger. This land, so bountiful and rich in biodiversity, historically has been used to produce things needed by other countries, while the local population continues to sink into the cycle of impoverishment.

April: On 30 March, Indigenous leaders demanded of President Óscar Berger and the Congress that they halt and cancel all licences for mining concessions. The Indigenous leaders and peasants are much better informed than the government officials, who continue to refer to mining projects as a factor that can help our country escape poverty. These officials can no longer claim to be unaware of the position of peasants and Indigenous leaders now that it has been expressed in public. Now we will see if they respect it, and take it into account, as President Berger promised they would.

Yesterday we met with the president. Thousands of us are worried about the possibility that Guatemalan territory will continue to be handed over for open-pit mineral extraction. We requested the suspension of concessions in most territories; that there be a technical and legal review of existing concessions; and that the current mining law also be reviewed.

August: In San Marcos the problems between the community and the Montana Mining Company are growing more serious every day.

Montana has sent out a flyer to the community that states, "The company's commitment is to work in a system that generates better distribution and water quality for your homes." This claim is just as false as one that says, "The Marlin mining project will not contaminate either the soil or the water." What are we to make of a company that states in the same flyer, "It is impossible for cyanide to make people ill"? This perversity knows no limits. The peasants of San Marcos are poorly prepared for the manipulation of the truth. It's possible that cyanide doesn't cause illness because it kills people!

September: Violence is once again spreading like an epidemic through our land. Protests against Canadian mining companies such as Goldcorp and Montana Mining, like those that occurred in San Marcos at the beginning of this month, will only get worse. People are finding that there is no way to get their needs and demands listened to. For President Berger and his circle, transnational mining companies are seen as a "blessing": they're not the ones who are going to end up without water. Not even the call to action of Monsignor Rodolfo Quezada Toruño, archbishop of Guatemala City, has given them pause. President Berger responded that the Guatemalan bishops are out of touch and that they oppose mining and hydroelectric projects without any justification. The deputy minister of energy and mines says that she is "ashamed to belong to their religion." The president has surrounded himself with people who support the surrender of our country; his team is totally identified with the interests of foreign investors. Yet Guatemalan bishops have spent years living with the people, understanding and sharing their pain.

November: Today, in the *Prensa Libre* newspaper, James Lambert, Canadian ambassador to Guatemala, wrote,

> Is it really possible for a country recognized as one of the most socially and environmentally responsible in the world, having one of the highest rankings in the Environmental Sustainability Index (ESI), to be at the same time an eminently mining country, whose mining industry contributes $41.1 billion to its economy? Why, yes, this is Canada! Like Guatemala, Canada is recognized around the world for its wealth in natural resources. The importance of natural resources to a country's development is undeniable. Nevertheless, it is only to the extent that one integrates social, economic and environmental considerations, taking into account the interests of all

affected sectors, that it will be possible to maximize the tremendous growth potential represented by the natural resources industry. This premise has as its central axis the concept of sustainability, an idea captured in the slogan of our Ministry of Natural Resources, "Canada's Natural Resources. For Today and for the Future!" In Canada, mining exploration and exploitation occurs in all of our provinces and territories, generating economic and social opportunity for many communities, including more than 200 Indigenous communities. Through sustainable development, these communities are creating the economic, cultural and social infrastructure for their futures and the futures of their children. Canada has always been an eminently mining country. Throughout a history of almost 150 years of mining production, we have become one of the most "intelligent" administrators, promoters and exporters of natural resources in the world. Today Canadian mining companies are on the cutting edge of advanced technology, environmental protection and social responsibility. That's why we're leading many of the most successful mining operations in the world.

It doesn't strike me as appropriate for Mr Lambert to compare a country like Canada to Guatemala. In the first place, let's look at size and population. Canada has more than nine million square kilometres of surface area; Guatemala has a little more than a 100,000 square kilometres. Canada (in 2001) has 31 million inhabitants; Guatemala (in 2002) has 11 million. This means that Canada has 3 inhabitants per square kilometre, whereas in Guatemala we have more than 100 inhabitants per square kilometre. Canada is in third place on the world's Human Development Index; Guatemala is number 121. Deforestation in Canada is -0.1% annually – in other words, the amount of forested land is actually *increasing* slightly every year; in Guatemala deforestation is at 1.0 per cent per year. In Canada the Indigenous population is 4.5 per cent; in Guatemala it's more than 50 per cent.

All this suggests that Canada is a large, sparsely populated country with a small Indigenous population, where the majority of the people live very comfortably. Guatemala is a small, densely populated country, with a large Indigenous population, where the majority live very poorly. I could go on, introducing information about justice and impunity, access to basic health care, etc., but I think I've made clear that the differences between the two countries are vast on many fronts and that comparison is not valid.

According to the United Nations, increased investment in extractive industries has had a large negative impact on the living standards of local communities around the world. It is this reality that Guatemalan communities that don't want mines appear to be confronting. I understand that one of the responsibilities of an ambassador is to support his country's foreign investments – since most of the transnational mining companies are Canadian – and that it is for this reason that Mr Lambert took the time to write an article on this subject. But I believe that it is up to us Guatemalans to make the relevant analysis of questions that affect our country and our future. The debate about mining should focus on the impact it will have, taking into account our national reality, and not by equating a country from the global North with one from the global South.

2005

January–February: The mining issue is at the centre of political debate right now. It is clear that no consultation took place in San Marcos, even though Montana Mining is exhibiting the signatures of peasants who were invited to different public lunches, then asked to sign to show their attendance. If this is the case – and it seems to be – then the government has violated the rule of law. This would mean that the granting of a licence to Montana is illegal and, therefore, should be declared null and void, as President Berger has been asked to do. Berger's unconditional support for Montana Mining's Marlin project in San Marcos is inexplicable, because the licence to this transnational was granted by ex-president Alfonso Portillo on 27 November 2003. Curiously, neither President Berger nor Juan Mario Dary, minister of the environment and natural resources, has expressed any doubts concerning the approval process for Marlin. My organization, the Colectivo Ecológico Madreselva (Madreselva Ecological Collective) has criticized Montana's Environmental Impact Assessment (EIA) on the basis of a technical review by Dr Robert Moran, a hydro-geologist with more than thirty-two years' experience in water and mining issues. It was to this expert that Montana denied permission to visit its mine after assuring the public, in a paid publicity campaign, that its project was open and transparent.

May–June: Fifty-one municipal deputies from the town of Comitancillo in the San Marcos region, led by the mayor and the city council, appeared before the deputy ministers of energy and mines, and of the environment and natural resources, and various members of Congress, to reject the mine. On various occasions they repeated, "We come here peacefully, doing things within the law." This is a gesture of confidence in our country's democratic system from the local authorities. The deputy minister of energy and mines recognized that mining can't take place without the consent of communities and local authorities. The communities of Comitancillo and their councillors gave a superb lesson in respecting the legal order and laid out a course that we can follow in our quest for a healthier, more just country.

According to a Vox Latina poll, published in *Prensa Libre*, 90 per cent of those consulted were of the view that there had been no official information to present the pros and cons of mining, nor had there been a consultation process. Nevertheless, Montana insists that sufficient information was provided. The result of the consultation that will take place on 18 June cannot legitimize a fait accompli: a licence was granted that violates the Indigenous peoples' rights to participate in a democratic process. Montana Mining, which until now has maintained that there was a consultation process, and that its project enjoys the support of the vast majority of the people in the Sipacapa region, is now making desperate efforts to halt this consultation. This demonstrates their fear of allowing the people to express their will. Montana presented an appeal on the basis of unconstitutionality before the judge of the Seventh Lower Court, who, with astonishing and unheard-of speed, allowed the appeal to be heard and formally suspended the consultation. The people of Sipacapa have decided to go ahead with their consultation process, because they have heard only rumours about this appeal. Municipal autonomy and the people's right to express their will are at stake.

July–September: Glamis Gold, Montana's parent company, affirms before the entire world that it has the support of the people of Sipacapa. The people of the region carried out a consultation process in good will, based on Article 169 of the Guatemalan Constitution, to expose Montana's lies. The Sipacapa consultation is a lesson in civil ethics, in pacifism and dignity, from the perspective of Indigenous people who are placing their faith in democracy.

The exploratory phase of the Marlin mine hasn't even begun and the first accident has already occurred. On Sunday, 25 September, a trailer transporting chemicals overturned while trying to cross a narrow bridge in the municipality of Malacatancito, Huehuetenango.

"The trailer doesn't belong to the company, but to another organization that is assisting the company," a Montana spokesperson explained.

October–December: In March of this year, a member of the Sipacapa community and I, on behalf of the Madreselva Ecological Collective, presented a complaint before the World Bank consultants who are looking into the environmental and social risks represented by the Marlin project. In September the World Bank issued a report in which it stated that the Sipacapa community had nothing to worry about. We became aware that Montana Mining is using photographs they took of us during our visit to the mine to tell the population that we – Bishop Álvaro Ramazzini, the leaders of the Sipacapa communities and the Madreselva Ecological Collective – are "collaborating" with them. This is absolutely false and makes it impossible for a dialogue to exist between us and them.

If Montana Mining, as the subsidiary of Canada's Glamis Gold, represents the new generation of mining, supposedly modernized, with cutting-edge technology, which acts in a responsible way towards people and their land, then we're screwed. The Environmental Impact Assessment used to approve the project was totally inadequate. Now there's a new EIA to approve a new mine, this one underground, in the same area as the first Marlin project. This new EIA does not mention the most significant potential impacts, or the long-term impacts, and for this reason it doesn't include mitigating measures. Since the EIA lacks precise data, the impacts of the new mine cannot be predicted.

EIAs have become a way of ridiculing the rights of the Guatemala people. The EIAs for the Marlin and La Hamaca mines in San Marcos and Cerro Blanco in Jutiapa have all been rammed through. They're all equally deficient. We've undertaken a partial review of the EIA for the proposed extraction project of Fénix Mining, located in El Estor, Izabal. This evaluation lacks the basic technical information required to evaluate the impacts of the mining process. In one of its first points it mentions that a separate EIA for a proposed nickel processing plant on the same site will be presented in the first trimester of 2006, making impossible an evaluation of the entire process.

Last week we had a meeting with Paul Wolfowitz, president of the World Bank. He was made aware of the violations of the rights of Indigenous people and the increasing environmental risks associated with the Marlin mine. Wolfowitz assured us that he understood that people were worried; he did not call us extremists or mindless enemies of development. He believes that things could have been handled better. He asked to be presented with an alternate proposal. It's up to people in Sipacapa to decide how they're going to respond to Mr Wolfowitz's proposal, for justice and the dignity of the people of the region are not negotiable.

The government has announced that, as the result of "official budget cutbacks," some communities in Sipacapa will have no schoolteachers in 2006. The solution suggested by government officials is that the communities accept schoolteachers chosen and paid for by Montana Mining. Could this be an official reprisal for the community's rejection of the mining project?

2006

January–March: People in Sipacapa are getting more and more desperate. It's already the end of January and there are no teachers in their schools. They refuse to accept teachers who will preach the virtues of gold mining to their children every day. The Ministry of Education says they must accept teachers recruited and paid by Montana Mining. The communities are divided now between the few who have found work in the mines and the majority who work on the land and wish to prevent the pollution of the mountains caused by open-pit mining.

The Ministry of the Environment has efficiently approved deficient EIAs for more new mines. The legal procedures for public consultation don't work. In Salem, in Sipacapa, the population has complained that the mining company has closed a road, adding half an hour to their walk to town. If they dare approach the mine, they are treated aggressively by Montana's heavily armed guards. On the government side, the High Commission on Mining has decided unilaterally not to suspend the new mining concessions now before Congress, even though this was one of the agreements signed by the commission. We thought it was worthwhile participating in this commission, in the search for peaceful solutions, but events have proved that those who believed that the commission was simply a government delaying tactic were right. Bishop Ramazzini has announced the breaking-off of dialogue within the commission because the government refused to implement the agreements that were reached. The minister and deputy minister of energy and mines assured the commission on various occasions that no new mining licences were being issued. Now we know that they were issuing new licences for uranium mining at the time they were making these statements. The mining industry's high-priced spokesman, President Berger, makes fun of people, whom he was elected to serve, who engage in peaceful protest. "I hope they don't get sunburn," the president said, "because then we'll need a commission to look into that."

What would you do, Mr President, if you lived in a town where 97 per cent of the population lived in poverty, if you had to live on the minimum wage of $200 per month and your government was authorizing operations that put your children's future in danger? People aren't marching to exercise, they're marching out of desperation!

April–May: Our weak institutions will end up rotting away out of corruption. The money and power of the transnational mining companies are enormous, while the interest of our public officials for Guatemala's future is nowhere evident. We could lose the fragile peace we've achieved because tension in the mining areas is increasing. We now know that in 2005 there were six new mining licences granted in four departments of the country, and there are 106 requests for licences in sixteen departments. All this took place while the High Commission on Mining was meeting to negotiate the bases for a new mining law. We don't know yet what's happened so far in 2006. What we do know is that Glamis Gold recently announced that its profits increased more than sevenfold in the last trimester, from US$2.2 million to US$16.9 million, due, above all, to the Marlin mine in Guatemala.

June–August: In Guatemala City, on Wednesday, 14 June, the Canadian ambassador, Kenneth Cook, reacted with surprise and amazement to the representatives of communities that feel they are negatively affected by Canadian mining projects. It became obvious that his own officials have kept him completely out of the loop on this file. The Guatemalan military also came along to "take care of" those Guatemalans who travelled from their communities to transmit their concerns to the Canadian ambassador. The dignified attitude of peaceful resistance of the people of Sipacapa remains intact, in spite of the mining company's blackmail and threats, and the humiliation the residents receive at the hands of the government of Guatemala. The company's policy of driving heavy vehicles through the centre of Sipacapa is a provocation that can only exacerbate the indignation of these people who have tried to defend their rights by peaceful means.

Our government is outdoing itself as a public relations agent for the transnational mining companies and the guardian of their interests. At the beginning of this year, the Guatemalan government owned 30 per cent of the shares in what used to be the Exmibal nickel extraction company. Today, thanks to the good offices of our officials, we own only 7 per cent of the shares. On examining the EIA for the Fénix project, we discovered that "on orders of the Deputy Minister of the Environment, Roxana Sobenes," the study had been split into two parts. This is a clear example of a public official whose management tactics flirt with illegality in order to secure the approval of high-risk projects. The EIA has other peculiarities that are worthy of mention: one of its most notable features is that its text of more than 600 pages is available only in English.

December: Ten years have passed since the signing of the Peace Accords that ended the civil war. Before, they killed people; the armed confrontation ended, but the majority of the causes of the war remain unaddressed. With the signing of the Peace Accords, conditions were created that allow the transnational companies to loot our remaining natural resources more efficiently. Guatemala's riches and its territory will be handed over by legal means to be exploited by local and foreign mining, hydroelectric, logging, sugar, and palm-oil companies. The assault on the natural world affects thousands of Guatemalans who need these lands for their survival.

2009

February: The EIA for the Marlin mine contains no information on the composition of the rock or details concerning the area's hydrology, hence Montana (now a subsidiary of Goldcorp since Glamis and Goldcorp merged) won't even have to justify any miscalculations. As for possible water contamination, the Ministry of the Environment can't even take samples from behind the dike unless the company authorizes it; we don't know whether they've built a holding tank for the decomposing cyanide or have discharged directly into the rivers. What we do know is that within the territory now controlled by Canadian mining companies, nobody may intervene. The recommendations of Archbishop Quezada Toruño and Bishop Ramazzini leave the new social democratic President Álvaro Colom, like his members of Congress, unmoved. The Indigenous communities in Guatemala are fighting for their lives. Through peaceful public consultations they have rejected mining on their lands. The current mining law favours the transnational companies. With only two amendments, our land and people could be defended: prohibiting open-pit mining and banning the use of cyanide. The mayor of Sipacapa has stated the Indigenous position clearly: "Even if they gave us 50 per cent of the royalties, we'd reject it. Here in the communities we're not interested in mining." Those of us who publicize information about the damage caused by extractive industries are called "subversives." I'm sorry! The established order is wrong here and everywhere. The established order wants to make us see the destruction of the only place in the universe where we can live as something normal and inevitable. The established order has trained us to believe that human laws are superior to cosmic laws and that progress, development, and the greed of the few are more important than life itself.

July–August: In order to import highly toxic chemicals, it's necessary to pay an environmental licence. Well, now it turns out that in 2005 Montana obtained the privilege to pay for this licence at a rate of only three quetzales (US$0.36) per kilogram of cyanide imported, instead of the legally mandated five quetzales ($0.60). It also turns out that Montana thought it was exonerated from paying for the environmental licence, so it didn't pay it. On the basis of the environmental licence to introduce cyanide into the country alone, Montana owes Guatemala more than US$2.5 million. This is a clear example of how mining companies that invest millions of dollars in propaganda in order to portray themselves as noble promoters of development refuse to pay the taxes they owe.

"There's been mining in Guatemala since the era of the Spanish conquest," is the mining companies' nagging refrain. Whenever a public official includes this line in a speech, it's a tip-off that he's been "influenced." Jonathan Salgado, an official in the Ministry of the Environment, recently used this line in his address to the International Conference on Mining in Guatemala City. He had the audacity to say that civil society is to blame for the fact that the government needs the tax revenue generated by mining, because the members of civil society refuse to pay their taxes, and it's the population's responsibility to contribute to the cost of the clean-up – as though that were possible! – that will be necessary once the mining ends. This, and other absurdities, made the audience wonder who this gentleman was really working for. It's offensive for Mr Salgado to try to justify mining in this way when it's public knowledge that the Canadian company Goldcorp – the only transnational that's mining gold in Guatemala at the moment – interpreted new laws as a total exemption from taxes, used its influence to reduce the import tax on cyanide, refuses to pay its debts even on this tax, and is now lobbying to have the import tax lowered even more.

October: In the lakes, rivers, and rainforests of the Atlantic Coast region of Izabal, the most critical issue is that of the land. Almost nobody knows how much of Guatemala's territory is planned to be sacrificed for the development of such a devastating, highly polluting industry as that of mineral mining, in such an important region as Izabal. This is one of the richest natural environments in the country, with the largest lake, the only Atlantic Coast, and unmatched biodiversity. It has enormous ecotourism potential and conditions that would permit the production of food and other agricultural products. But Izabal is also socially fragile and hosts some of the poorest communities in Guatemala. Land abandoned years ago by Exmibal was reoccupied over the last few decades by landless Guatemalan families. Since the signing of the Peace Accords, negotiations with government institutions had begun to legalize this informal land tenancy. With the rise in the price of nickel, transnational nickel mining companies – led by INCO from Canada, also – have entered the region to grab whatever they can. These powerful companies are using all of the resources at their disposal to kick the peasants off the land, or, when this fails, to accuse them of being invaders. Some peasants have already died in these conflicts. It's easy to imagine the anger and frustration these Guatemalans must feel at not having even the most limited of their rights guaranteed, while "their" government betrays them to get on the good side of the mining companies. They are, literally, playing with dynamite.

September–December: The Indigenous peoples thought that President Colom, who campaigned as the representative of their interests, would halt the advance of the mining companies onto their territory, or at least that he would try. But so far – nothing. Engineers from the United States have presented a report, after years of monitoring the situation, in which they conclude that the cracks in the walls of the houses around the Marlin mine are related to the repeated detonations and heavy-vehicle traffic from mining activity. "We're going to initiate a serious study to prove that the damage to residents' houses is not Montana's responsibility," the company's representative stated. They prefer to spend thousands of dollars on "serious studies" rather than compensate people for the damage they're causing, just as they prefer spending money on lawyers rather than paying the taxes they owe.

Engineers had to come from the United States to prove that the origin of the cracks in the walls and floors of the houses are the detonations to smash the rock and the constant passage of gigantic trucks that carry away sections of the mountainside to be ground down in the mills. The local representatives of Goldcorp in San Marcos explained that the cause of the cracks in the houses was that the residents were playing their music too loudly. Here we need foreign experts to prove the mining companies' misdeeds; here, our government is unable to protect its people; here, as elsewhere in the world, modern mining is a process that destroys, sickens, impoverishes, and corrupts.

2010

January: The Supreme Court of Canada has ruled that large mining projects must have a comprehensive Environmental Impact Assessment, without dividing the project into different slices, and one that guarantees public participation. The Supreme Court decided that the Canadian government violated applicable norms. The judgment states that, in the evaluation of the Red Chris mining project – a vast open-pit gold and copper mine in northwestern British Columbia – the project was fragmented illegally, making it impossible to know the true environmental impact, explained a communiqué from the Asociación Interamericana para la Defensa del Ambiente (AIDA; Inter-American Association for Environmental Defence). This is happening right now in Canada, a country with a long mining tradition, and one whose institutions work far better than do ours. "Throughout the hemisphere we have witnessed countless projects with immense environmental and social impacts which unfortunately are presented and evaluated in parts, open-pit mines are a recurrent example of this, for which reason this judgment is vital for the entire region," the director of AIDA stated. This is a fine example of how mining companies "bamboozle" the authorities, even in countries like Canada. One recommendation made to numerous Guatemalan officials by Robert Moran, a hydrologist who has studied the EIA for the Marlin mine, is that Guatemala must demand a cumulative evaluation of the impact of all the mines that Goldcorp plans to exploit in the San Marcos region. Through its subsidiaries, Montana and Entre Mares, this transnational has more than ten "mining interests" in the northwest highlands alone, each of them contributing its quota of environmental destruction, high levels of water consumption, the release of toxic chemicals, and the creation of social conflict. They will all affect the same territory; taken together, their impacts are much more dangerous. Up to this point, so far as I'm aware, no such study exists of any of the regions of "mining interest." What does exist, in an echo of the Red Chris mine in Canada, is a fragmented EIA, authorized when Juan Mario Dary and Roxana Sobenes were minister and deputy minister of the environment respectively, for nickel exploitation in Izabal. That a national Supreme Court has condemned this practice establishes an important precedent. This is a crucial moment for our future: a new mining law is under discussion.

Will the authorities cancel mining authorizations made on the basis of fragmented EIAs?

June: The news that the government is going to carry out cautionary measures recommended by the Inter-American Commission on Human Rights (IACHR) of the Organization of American States (OAS) has prompted mixed reactions. President Colom is bending to the orders of an international body, but the government's responses to the IACHR suggest that it does not accept the pollution of the environment, the impact on community health, or the persecution of community leaders as problems caused by the Marlin mine. Nor will the mine be closed; there is simply a temporary suspension of mining activities. It was necessary for an international body to take this step because the petitions and entreaties of thousands of Guatemalan citizens had received no reply at all. The president did the right thing in fulfilling the order to suspend mining activities, but this is not enough; he must implement all of the measures recommended by the IACHR.

July–August: This week the Ministries of Communications, Energy and Mines, and the Environment presented their joint report on the cracked houses around the Marlin mine. According to the three ministries, the investigation, which lasted five months, has concluded that the cracks "have no relation to gold exploration in the area." This surprising find contradicts a similar report, undertaken by a group of professional engineers, and including geologists from the University of Colorado and a civil engineer from the University of Pennsylvania, and a mining engineer who has more than forty years' experience with mining companies and the state government of Colorado, as well as a Guatemalan professional. This investigation, which lasted two years, originated in the initial complaints in 2007. The team observed numerous cracked houses around the mine and, as a control sample, similar houses in villages that were far from the mine. They conclude, "The most probable cause of the cracks is vibrations in the earth. There are no sources of vibrations in this area, except for those derived from the mine's explosions and the passage of heavy trucks." Their report is online and publicly available; the other report – the official one – has been seen by nobody.

A nun from San Marcos asked, "If there's no relation between the cracks and the mine, why didn't the houses crack before, and why did they start to crack when the mining work began?" The official report says that the causes of the cracks are poor construction techniques and seismic activity. If this is the case, why don't all the houses in San Marcos have cracks, only those close to the mine?

In mid-August E-Tech International, commissioned by Oxfam America, presented its evaluation of water quality at the Marlin mine. I will cite some of their conclusions: "The EIA provided limited information ... the monitoring period for the baseline water quality was too brief ... no subterranean water samples were taken from deep in the site ... the possibility of seepage from the tailing reservoir was not considered ... almost half of the residual rock has the potential to generate acids." These are the same conclusions that were reached by Dr Robert Moran's World Bank–financed study in 2005. The IACHR, the analyses of two prominent Guatemalan research institutes, and the report of Doctors for Human Rights all indicate the problems and dangers associated with the Marlin mine. The government and the mining companies ignore all of these reports, grabbing all that they can, day and night, without ever stopping, while spending millions in stupid, offensive propaganda. Those adversely affected by the mine are told that they must prove "scientifically" its negative impact. Communities that barely eat must seek financing and training to understand, track, and defend themselves from these dangerous operations. I feel myself sinking into deep distress as I grasp the impossibility of preventing the government imposition of oil drilling and mineral mining projects by means of analysis or scientific studies in the conditions of corruption, ignorance, and injustice in which we find ourselves.

October–November: Montana, the subsidiary of the powerful Canadian corporation Goldcorp, has just given us a master class in lying and deceit. It's worth analysing the advertisements they paid for. "Marlin undertakes the discharge of waters from its mines in the presence of regulatory institutions." In other words, Montana pours contaminated waters into a gorge in the basin of the Cuilco River. The advertisement assures us, "Montana fulfilled the commitments established in its EIA and by the rules concerning the discharge and re-use of residual waters and the disposal of mud. The discharge occurred in a framework of transparency. The public regulatory institutions undertook monitoring and supervision."

Lies. Montana did not warn the Ministry of the Environment before-hand and did not have formal permission to pour contaminated water into the river; hence the pertinent government regulatory institution was unable to adequately supervise this activity. By coincidence, a member of the ministry's monitoring team happened to be in the area and took samples – wait for it – a few hours *after* the discharge had occurred. "In its majority, the discharge consisted of excess rainwa-ter that had accumulated in the tailing dam during the rainy season and which, prior to being discharged, had been filtered through the region's most modern water treatment plant," the announcement says. It doesn't say that what is contained by the tailing dam is a danger-ous, contaminating toxic soup. In addition to water, the dam contains crushed rock, all of the chemicals used in the mining process, includ-ing cyanide, explosives, oils, greases, fuels and anti-coagulants, waters from the drains of the industrial complex, the laboratories and other operations, and possibly fertilizers and pesticides. We Guatemalans don't know what's in the crushed rock, since this information was excluded from the EIA, so that we can only speculate; but it could include particles of aluminium, antimony, arsenic, barium, cadmium, copper, chromium, cobalt, iron, lead, magnesium, mercury, molybde-num, nickel, selenium, thallium, titanium, tungsten, vanadium, zinc, and radioactive elements such as uranium and thorium originating in the rock itself; magnesium sulphate, nitrates, ammonium, boron, phos-phates, and chlorine. These materials mix and react; very small doses of some of these elements could cause significant damage. Montana did not advise the government of this discharge and – showing full trans-parency! – discharged at night. They admitted to having discharged only when their work was done.

Goldcorp's cynical propaganda claims to be creating progress and well-being for Guatemala while the economic study of the highly respected centrist think-tank Asociación de Investigación y Estudios Sociales (ASIES; Association for Research and Social Studies) demon-strates that, with the arrival of gold mining, Guatemala has lost much of what it had gained. The mining regions have been plunged into conflict, community leaders are persecuted, houses are cracked, the water table is being depleted, and Canadian transnational companies have taken possession of communal land by illegal means. According to an unpublished report by the Prospectors and Developers Association of Canada, founded in Ontario in 1932, "Canadian mining companies are four times as likely as their competitors from the rest of the planet to be implicated in violations of the principles of corporate social responsibility." And no one in any of the government agencies whose task is to monitor mining has sufficient knowledge of the subject to adequately protect the interests of Guatemala and its people.

2011

January–February: Goldcorp has reported its earnings mining gold and silver in Guatemala over the last five years at more than US$2 billion. They are not paying any other taxes, so you would think they might pay the Guatemalan government the puny US$22 million they owe for importing cyanide. Goldcorp's CEO has told the press, "We're working on it." The temporary closure of the Marlin mine, ordered by the IACHR, has simply allowed Montana to wear down the resistance of some of the community leaders who opposed the destruction of their land. The frustration of the Indigenous communities at the violation of their right to be consulted and the utter compliance of both the conservative President Berger and the social democratic President Colom unleashed a social conflict that has left entire communities mentally exhausted, and unfortunately some of their leaders have given up. Now Goldcorp is boasting that on 2 December 2010 Vice-President Rafael Espada convened round-table discussions among interested parties in San Miguel Ixtahuacán and Sipacapa to "ensure that the mining industry develops a standard of respect for communities and the environment that will benefit the country." Ha! According to a Goldcorp press release, "This marks the beginning of a participatory process among the national government, the local municipalities and Montana." What about the people who have already been affected, with cracked houses, sick children, and trumped-up law suits brought against them by the mining company? It's at their expense that the "participation" of a few local mayors has been achieved.

July: It's been seven years since the rights of the residents of Sipacapa were violated by the concession of a mining licence on their communal land, without their being informed or consulted, as is required for an Indigenous community. But nothing has happened. Water sources have dried up, skin and lung diseases are rampant, houses are cracked, and no government institution has been able to help these people; on the contrary, government agencies deny the truth of the residents' claims. By contrast, Montana enjoys constant protection from armed guards. As soon as the company complained that eight "dangerous women" had taken down an electric cable, belonging to the mining company, which crossed the land of one of these women, judicial orders were issued for their arrest, along with that of several leaders who had organized peaceful protests against mining activity. All were promptly detained. Swift, efficient institutional responses in favour of the mining company have been a constant since the beginning, when President Berger dispatched more than a thousand Guatemalan soldiers and police officers to escort the mining cylinder into the region. There is no independent judiciary in Guatemala; what we have is a system that promotes impunity.

Over the last six years, more than fifty-three Indigenous communities have held consultations, in a clear demonstration of their democratic culture and capacity for participating as citizens, to express their unequivocal rejection of mining on the lands they inhabit. More than one million adults have participated in these democratic exercises, yet it is clear that the majority of the presidential candidates attribute no importance to the position of such a large number of their compatriots. They live in such different worlds that they fail to understand that what they consider "development" has no meaning for Indigenous peoples, who see the earth as their mother and not as a marketable good. Or are they afraid of scaring away big finance? One candidate said that he didn't want to put at risk the rights of transnational corporations. What about the rights of the Guatemalan people?

September: Here in Guatemala, overwhelmed by presidential campaigns full of empty promises, we hear nothing. Maude Barlow of the Council of Canadians arrived in Guatemala to "check out what mining companies of my country are doing to countries like yours."

Ms Barlow said, "The mining companies are destroying Canada's international reputation. The federal government has promoted and supported the expansion of Canadian mining without any consideration for human rights or environmental impact." Ms Barlow asked that the Canada Pension Plan, which holds $256 million of Goldcorp shares, demand the mine's closure and that the Canadian government help to clean up the site. She pointed out that the Declaration of the Rights of Indigenous Peoples, to which Canada is a signatory, requires participating countries to ensure that their countries' companies respect the right to prior, informed consent before mining operations can take place on Indigenous land. "What moved me most deeply was the courage of those who are willing to risk their lives to maintain their resistance, against all odds, in search of their right to water," Ms Barlow said, prior to leaving.

Guatemalans did not hear about Ms Barlow's visit to our country. Similarly, in December 2010 Goldcorp boasted publicly at having been admitted to the Dow Jones Sustainability Index. But in September 2011, when they were kicked out of this index, there wasn't even a press release to announce this event. Why was Goldcorp kicked out? Surely the Dow Jones Stock Exchange is run by "eco-hysterics" and "eco-terrorists"! A recent report by Tufts University concludes that the Guatemalan government's revenue from the high profits of the mining companies is negligible, and far below global norms; that the environmental risks of mining as practised in Guatemala are extremely high, and will get worse after the mines are abandoned; and that the Marlin mine contributes very little to Guatemala's development.

October: The new president, retired General Otto Pérez Molina, has named Roxana Sobenes to the post of minister of the environment. She was deputy minister to Juan Mario Dary in the government of President Berger. She was the one who had the brilliant idea of dividing Environmental Impact Assessments into slices. This wouldn't have been so serious had she not been a public official charged with the obligation of overseeing our natural heritage. Her other recognized feat was as an independent consultant, when she was part of the team that wrote the EIA for Tikal Minerals, a company that wishes to mine minerals on the beaches of the Pacific coast. Why did Pérez Molina name Sobenes minister of the environment? Maybe he doesn't know about her connections with oil and mining companies; or maybe, in appointing her, he is repaying a debt for the financing of his campaign. The one thing that's certain is that she was not named for her merits as a defender of the environment. Maybe I'm wrong; it's happened to me before. Let's hope so! I will be ready to apologize. I hope that our new president has the wisdom to understand that taking care of our house is essential to our security.

2012

March: The Pérez Molina government has launched a media campaign of defamation and intimidation to smear and decimate, at any cost, the defence of Guatemalan territory before the advance of transnational oil and mining companies. Many people fear that we could return to a period of repression and violent conflict between our society's wealthiest and poorest sectors. My hope is that social media will prevent the mass media that are in the hands of the powerful from succeeding in characterizing anyone who defends his land from forced occupation by powerful transnationals as a terrorist. Today I would like to draw attention to the thousands and thousands of women who are putting their lives at risk to prevent the destruction of Mother Earth. For all that the media portray them as criminals, these brave women are defending life. I remember the words of Father Antonio de Montesinos in the sixteenth century, when the Spaniards took possession of all of the wealth they encountered upon their arrival in the New World: "Are these not men? Do they not have souls? Are you not obliged to love them as you love yourselves? Don't you understand this? Don't you feel this? How can you fall into such a deep sleep?"

That priest's harangue, to a throne whose greed knew no limits, may be compared to the valiant posture of those who dare to defy the power of the mining companies in order to defend the most vulnerable sector of Guatemalan society: Indigenous women. They march under the sun, day after day, in the hope that their gesture will move us. They don't know that here in the capital they're portrayed as poor, ignorant, and easily manipulated by international activists. Please! Before we swallow that convenient caricature painted by certain "communicators," let's think about what it means for these women to leave the security of their homes and their daily labours, which give them the food they eat, to walk for more than a week without knowing what will happen. These are Guatemala's most vulnerable, excluded, and needy people. Among those who are marching today, I would like – again – to highlight the role of women. Their role in this march is striking. In San Marcos, while some leaders succumb to temptation, beneath the pressures of frustration and government blackmail, it is the women who, in spite of persecution and even bullets, resist and defend their land, their dignity, and the future of their children. One of the demands that the peasant march is making of President Pérez Molina's government is that it implement the moratorium on mining licences. The technical justification that has been used by the mining companies' spokespeople, the members of Congress who are friends of the mining companies and other government officials, is that the Constitution of the Republic defines mining as being of public utility. Our Constitution was written in the last century, when mineral extraction did not constitute an environmental and social problem of the dimensions it presents in the twenty-first century. We need to ask whether, in our present circumstances, this Article remains valid – especially since this same Constitution also contains Article 97, on ecological balance, which states, "The State, the municipalities and the inhabitants of the national territory are obliged to supply the necessary social, economic and technological development to prevent the pollution of the environment and maintain ecological equilibrium."

Are you listening? You are *obliged* to *prevent* pollution. No matter how much money Goldcorp invests in propaganda, the reality is evident to anyone who visits the Marlin mine. The "ecological equilibrium" of the region is broken. It's not possible to take out the 36 million tonnes – according to the work of science fiction known as the "Environmental Impact Assessment" – of crushed rock that remain after extracting the valuable minerals, mixed with toxic elements and metals that are dangerous to human health, without ruining the environmental equilibrium. Now that they have destroyed this mountain they are going to continue exploiting its entrails by way of tunnels; then they will move on to other areas. No doubt they will cover the evidence with something green, such as a so-called reforestation plan. And there it will remain forever: the evidence of what mining leaves behind, in addition to the social conflict, divided communities, health problems, prostitution, and violence that now characterize San Marcos. The Indigenous peoples and peasants who wish to avoid having this happen in their communities are fulfilling Article 97 of the Constitution of the Republic.

June: This Wednesday a woman who was protesting gold mining was shot in the back. I'm outraged by the cowardice of certain sectors; I'm sick of writing about mining when every argument loses. By introducing state-of-siege legislation and militarizing the mining sites, President Pérez Molina's government has signalled that it has no hesitation in using the institutional violence of the state against the people. We demand justice, but we know that the justice system is occupied territory. Acts such as this risk serving as catalysts. My respect and solidarity for Yolanda Oquelí, heroine and survivor.

July–August: In March I received an invitation to participate in an international tribunal that will be held in San Miguel Ixtahuacán to identify possible damage to human health caused by the mines of the Canadian transnational Goldcorp in Guatemala, Honduras. and Mexico. On 5 June I sent a request to the CEO of Goldcorp-Guatemala, Mr Mario Marroquín, requesting permission to visit the Marlin mine in the company of Robert Goodland, who worked for twenty-three years as an environmental evaluator for the World Bank, and who was also attending the tribunal. Mr Marroquín replied on 19 June, saying, "Due to a busy working agenda in July and difficult climactic conditions, it is not possible to concretize this visit." How surprising that Goldcorp can forecast the climate a month in advance! One wonders whether this refusal has anything to do with the fact that the tailing dam is on the point of collapse. The initiative for a tribunal on health originated when a nurse, a graduate of the University of Western Ontario, undertook a study in San Miguel in 2011, whose preliminary results indicated that mining operations had increased the community's sociopolitical vulnerability and threatened the well-being of local residents. Reports of insomnia and other sleep ailments, extreme anxiety, and desperation show that the mining operations have affected residents' health. These results meet the prerequisites established by the Ottawa Charter for Health Promotion; for this reason, numerous Guatemalan organizations have convened the tribunal to define the damage to the health of the three communities most severely affected. So far we have spoken only of damage to the environment and physical health; it's important to recognize that damage to mental health can cause other types of long-term problems for which we lack preparation and, even more so, the training to treat them.

The visit of the president and vice-president of Goldcorp to Guatemala is not innocent. According to an email sent to Mining Watch Canada on Monday, Goldcorp Vice-President Brent Bergeron and Goldcorp President Ian Telfer will "make a fascinating journey" to Guatemala, in the company of "four Canadian members of Parliament." (In fact, the attendance of only two MPs has been confirmed: Conservatives Dean Allison and Dave Van Kesteren, both of whom sit on the Foreign Affairs Committee. Their attendance was organized by former Liberal MP Don Boudria, now working as a Goldcorp lobbyist; but no Liberal or New Democratic Party MPs have accepted the invitation.) The delegation will travel in "Goldcorp aircraft." In the capital they will "meet with government ministers" before returning to Canada. Six months ago,

Bergeron testified before the Standing Committee on Foreign Relations and International Development of the Canadian Parliament: "In Guatemala, I would like to see them modernize their mining regulations. That would add to the stability of the environment within which we deal in Guatemala. Can I go in as Goldcorp and start training the Ministry of Energy and Mines? I can't do that. The credibility behind that is not right. However, I think it makes a lot of sense to have a government institution come in to take our experience here in Canada – the National Resources Canada in terms of their experience – and bring that experience to Guatemala."

The social confrontations that were unleashed once it became clear that the mining model was being imposed on our country are becoming more severe. The opposition of millions of Guatemalans doesn't matter a damn to President Pérez Molina, as it didn't matter to Berger or Colom, because the presidents are direct beneficiaries of this activity. Yet I sense a dangerous change of outlook among thoughtful people in the population. As a result of the incessant public relations bombardment in favour of mining, many reasonable citizens now entertain the possibility that large-scale mining, carried out in a sustainable and reasonable way, may be possible. Would it be ethical for a Goldcorp executive to come and propose changes to Guatemalan legislation? What does it matter, if ethics, transparency, legality, and justice don't matter a damn either to our presidents?

September–October: Yolanda Oquelí addressed the people who gathered opposite the entrance to the mine last Sunday to tell her comrades in the anti-mining struggle that she is aware of the rumours that are circulating about her. Anonymous sources have been telling people that the attack that nearly cost her life, leaving her with a bullet lodged very close to her spinal column, was faked and that in fact Yoli received several million dollars and went away to Canada. Maybe they fomented these rumours about her because they thought that this tiny woman would never dare to return to the spot where she was ambushed. Violence, fear, deception, and lies are the chosen weapons of those who impose mining on us. (On my last visit to San Miguel Ixtahuacán, I was stopped in the street by people who were angry with me because they had been told that I had stopped fighting the mining companies and joined them.)

I had never been to San José de Golfo. Last Sunday I was invited by two women from the community and witnessed the return of one of the bravest women I've met. Yolanda Oquelí's iron will was clear as she told her comrades in the struggle, in dignified tones, that they must remain united and must not allow themselves to be provoked or tricked. It's very moving to watch her walk, her pain evident even when she's smiling. On seeing the reality of the people of San José, who have spent seven months camping on the highway in shifts to stop a project that threatens their children's future, in an admirable gesture of peaceful self-defence, I realized that the differences between them and those who are desperate for gold are unbridgeable. I challenge the defenders of gold mining to explain to us what use this metal has in Guatemalans' lives, and how its use makes our lives better.

The new mining law, which was finalized during the visit of Golcorp executives and shareholders, and two Canadian members of Parliament, in Goldcorp's private plane, was described by Óscar Clemente Marroquín, editor of the newspaper *La Hora*, as "abject, cynical, rotten and contemptible." According to Goldcorp, Guatemala's new mining law was discussed with "all interested parties," though those of us in the general public still don't know what it contains. It appears that articles relating to environmental standards and the obligation to consult affected communities have been abolished. The Congress enjoys neither the people's confidence nor the technical capacity to assess the law. Only President Pérez Molina can stop this. But will he?

December: Years ago I was told that one of the most terrible effects of mineral mining was the divisions it created between people. I thought this was an exaggeration. I couldn't conceive of anything worse than environmental destruction and pollution. Today I understand better, though I still see differences: while it is impossible to remake a mountain and restore potable water to a mined area, where acid drainage may flow forever, divisions between people can be healed.

As though we weren't already sufficiently divided in Guatemala, mining is creating a new abyss, as deep as a mine shaft, between those who don't want it and those who are imposing it by force. If we count those who want mining as including local collaborators (industrialists with mining connections, workers, government officials, mayors, and members of Congress) as well as a few columnists and bloggers who attack Indigenous people and ecologists at every opportunity, we can see that they are relatively few. If we observe the brazen defence of the mining companies made by "our" government when they send in riot police against women "armed with religious symbols" (this is how they express it), with which they hope to halt projects plagued by illegalities, we can see the grotesque division between those who believe in brute force, and those who have only their courage and convictions. The image of more than one hundred black trucks, full of heavily armed police – as though they were off to combat the most dangerous of delinquents – who use tear-gas grenades to dislodge – without any judicial order to do so! – people from roads clearly illustrates that "our" government is ready to use force against the people who elected it, including the elderly and small children. If those who, influenced by "official" propaganda, support the imposition of mining, took the trouble to travel to La Puya, where this occurred, which lies only twenty-nine kilometres from the capital, I'm sure they would understand (provided they don't have an economic stake in a mine, which seems to blind everyone) that these courageous people are prepared to lose their lives before they lose their children's water.

2013

February: Last year the IACHR ordered the Guatemalan government to provide Yolanda Oquelí with protection. On 7 December, nine months after residents blocked access to the entrance to the mine, the government sent more than seventy patrol cars full of armed men to dislodge the residents and open a corridor for the mine's employees. The anti-riot police ran into a human barrier and a woman who asked to see their judicial order for removing the protesters. By order of the deputy minister of the interior, Yolanda was without the bodyguards who had accompanied her since the implementation of the IACHR order. It must have been difficult for those men to confront a woman who showed no fear of them and maintained calm among her comrades, who were lying on the ground singing hymns. They, armed to the teeth, arrived without the legal authority to fulfil their orders. Yolanda's exemplary reaction must have been caused by anger, as a government official had decided to leave her without the bodyguards mandated by the IACHR, until the Ministerio Público (MP; Public Prosecutor's Office) intervened. According to an analysis by the Presidential Human Rights Commission, Yolanda Oquelí and her family are very vulnerable; it was recommended to increase the level of protection provided by the government. There was resistance, as the government argued that the state did not have the resources to provide this protection. This argument is ridiculous when the same officials who are making it decide to send dozens of patrol cars for hundreds of kilometres to protect the interests of a transnational mining company. Now the government wants Yolanda to sign an agreement that her bodyguards will rotate every thirty days. The justification for this rotation is the bodyguards' security! They are at risk if they spend too much time with her! Can we really imagine the pain and traumas that Yolanda and her family have suffered? A bullet that will remain lodged in her fragile body. The constant, necessary presence of security personnel. The arrogance and violence with which she is treated by the miners and the very government officials who are obliged to protect her. Yolanda Oquelí has not committed any crime; she is an exemplary citizen who has given us a lesson in peaceful resistance. She deserves respect and she deserves protection. If anything happens to her, the president and those who follow his orders will be responsible.

Tomorrow, 28 February, it will be one year since the resistance began in La Puya. Local women have worked in shifts, twenty-four hours a day, 365 days a year, to block the entrance – not the highway, not the road – to prevent access to the El Tambor mine; to a site where a mining company plans to drill, with the support of the government and legal chicanery. If necessary, the transnational mining company will blow the mountain to pieces. After carting away everything that has market value, the miners will leave behind heaps of debris – thousands of tons of mercury, copper, selenium, or arsenopyrite with a high potential to pollute and damage human health; these elements will be mixed with toxic materials used by the metallurgic industry, such as cyanide. What would you do if this were happening in your backyard? Would you allow it?

Would the irreversible damage occasioned by gold mining be justified if there were a real economic benefit for the local people? Because it's clear that not even the pitiful 1% of the profits that's mandated by law will remain in the country given that no one is calculating the amounts. For the region's inhabitants, economic benefit would not be an acceptable justification given that mining puts their water at risk. The greatest fear mining causes them – and it is justified – is the overuse of water. Those who live in Guatemala's "Dry Corridor" won't have time to get ill from chemical contamination because long before that they will die of hunger or thirst. Climate change is the biggest crisis of our era. The conditions of life are already much more difficult for the majority of the Guatemalan population that depends on its harvests to survive.

April: The approval of two new mining projects – proudly touted by the government as the first projects authorized by the new president – has seen Pérez Molina give away more Guatemalan territory to Tahoe Resources. In numerous public consultations, the people of the nearby community of Santa Rosa said that they did not want mineral mining. The EIA for the new El Escobal mine is a farce; no decent Ministry of the Environment would have approved it. Tahoe Resources may now be the owner of El Escobal, but Goldcorp owns 40 per cent of its shares; Kevin McArthur, the founder of Tahoe, is a former Goldcorp president and CEO. The residents of Santa Rosa can visualize their future by looking at how Goldcorp acts in San Marcos. The adversary is terrible; but it's more terrible to confirm that no one is capable of holding back the greed that unites those who are ready to destroy our land with explosives in return for a few ounces of metal.

Rest in Peace, Lake Izabal.

As the Pérez Molina government imposes state-of-siege legislation, militarizing numerous mining communities, Guatemala is becoming the laughing stock of the international judicial community. Guatemala has already been condemned by the IACHR as a result of the Myrna Mack case (described in chapter 5 by Helen Mack Chang). Now our supreme judicial forum, the heavenly Constitutional Court, is changing laws in ways that can be replicated in any future criminal proceeding. In order to declare a state-of-siege in Jalapa, Mataquescuintla, Casillas, and San Rafael Las Flores, the judiciary and the government worked as a team to foster the impression that rather than citizens who oppose mining, these regions are full of criminal gangs who killed a policeman. The pretext used to justify sending the army to these areas collapsed when the MP released recordings of the San Rafael mine's security chief speaking to its public relations consultant. Here are some of the highlights:

> "Just kill the sons of bitches!"
> "Goddamned dogs don't realize that mining creates jobs."
> "If I don't want trouble with the law, I'm gonna have to leave Guatemala for a while after this job."
> "We can't allow the resistance to get started, another La Puya–no way!"
> "We've got to get those piece-of-shit animals off the road."

The Ministry of the Interior assures us that firearms were not used against the population. A pseudo-program on the pro-mining television channel claims that local people attacked poor unarmed miners. We know that there are good and bad people everywhere, but it is unjust that the government locked up innocent citizens in a maximum security prison, where they spent several nights, and accused them, and the priests who tried to prevent a bloodbath, of being terrorists.

Today the state-of-siege legislation is being removed. But can lost confidence in government be regained?

July: Amid clouds of smoke and open deceit, things are happening in San Rafael that should not go unnoticed. The escalation began when local residents declared their opposition to mining on their land. At the end of last year, someone incinerated a store and a police patrol car. Explosives and detonators were removed from Tahoe trucks. At the beginning of this year, highways were blocked and two guards from a private security company were reported dead. Tahoe's security guards opened fire on community members who were standing in an area away from the mine's property. The Ministry of the Interior did not hesitate to indict local residents and leaders for these acts of violence.

As a result of this confusion, the MP tapped the phones of key actors. In this way we learned that Alberto Rotondo, the mine's security chief, ordered the security guards to open fire on the civilian population. The MP also discovered the espionage network manipulated by the mine's security department, which involved – among other surprising characters – members of the National Police. This espionage apparatus, which was set up to track the movements of community leaders who opposed mineral mining, infiltrated "social leaders" into the resistance movement, including the "activist" who is now on trial for killing a police officer.

If not for the efficient intervention of the MP, laying bare Tahoe's tactics, innocent citizens certainly would have felt "the full weight of the law." Now that the First Court of Appeal has agreed to hear a complaint against the San Rafael mine in El Escobal (which suspends the mining licence and prohibits the resumption of mining activity), we know that the wall of impunity that has surrounded mining can be broken. We can dream of applying the full weight of the law to those who violate it.

October–November: Between 2011 and 2012, in spite of all the boasting of its public relations personnel, production from mines and quarries accounted for barely 3 per cent of Guatemala's total GNP. In an excellent article published in *Plaza Pública*, Jonathan Menkos asks, "What economic interest can there be in mining for us to put this production ahead of democratic governance and the remaining 97% of national production?" Another fact from Menkos: the exorbitant taxes paid by the mining companies account for barely 1.1 per cent of the government's total tax revenue (in a country where tax-collection is notably inefficient and many people fail to pay taxes).

A Tahoe subsidiary is opposing the San Rafael closure order in court, arguing that local residents cannot rescind a mining licence that was already issued. Yes, it was issued – but illegally, the day before local people were informed of the mine! This episode gives us an idea of how mining corrupts public officials. Our officials are no longer loyal to our country. That much, at least, is clear. Sixty-three per cent of Guatemalans oppose all mining, yet more than one hundred mining licences are now operative.

What makes Guatemala attractive to mining companies – because there is gold in lots of other parts of the world, for example in Canada, where most of these companies come from – is that here there is neither the political will nor the administrative capacity to adequately supervise mining activities. Nor do companies have to worry about the legal consequences of accidents or abuses committed by security personnel. EIAs that would be unacceptable in any developed country are approved in Guatemala. The government never fulfils its duty to consult the population, particularly not when dealing with Indigenous people. Nor are cost-benefit analyses performed to discover whether Guatemala has the conditions necessary to exploit its mineral resources in a way that would be beneficial to the population. As soon as negative impacts become evident, the government and the mining companies simply deny everything. From one day to the next, serious environmental and health problems appear, but by the time their effects grow severe, the mining companies are no longer there. All that remains is pollution, destruction, piles of debris, ill communities, and a Guatemalan government that has cynically ignored the concerns of its population, terrorized and defenceless before one of the most dangerous industries of modern Western civilization. Guatemalan society, indifferent to the conflict that is developing before its eyes, prefers to remain silent.

Final Testimony: April 2016

I was a columnist for the newspaper *Prensa Libre* from 1996 until November 2015. I used the space offered to me by Guatemala's largest-circulation daily newspaper as a tool to carry out my work as an ecological activist. From 2003 onwards, I devoted nearly all of my columns to modern metal mining, and I gave space to the affected populations who chose peaceful resistance to the surrender of their land to foreign transnational corporations.

My stance is radical and uncompromising: Guatemala does not enjoy the conditions necessary for the modern metal mining industry to provide benefits to the Guatemalan population. Institutional weakness and corruption make this impossible. The vast profits produced by the exploitation of precious metals go to the foreign companies' local partners, to the politicians involved in making the mineral exploitation possible, and to those workers who obtain short-term employment. The losses, in environmental, social, and economic terms, are enormous, and in many cases irreversible. We've already lost enough clean water, woods, soil, health, community harmony, and social peace. But the worst part of the exploitation of precious metals in Guatemala is the way in which it affects relations between governments, peoples, and our precarious justice system.

In 2016 the mining of precious metals continues to generate social and political conflict in Guatemala. The population is aware of the high levels of corruption in recent governments and associates them clearly with mining companies. In the protest marches that overthrew the government of Otto Pérez Molina in 2015, precious metal mining appeared as an important theme. More than 90 per cent of the population now opposes metal mining; but this position of citizens has been cynically ignored by every president since 2005. The current incumbent, President Jimmy Morales, is no exception. Señor Morales's government has already shown so many signs of incompetence, and of a lack of any clear policies, that many analysts doubt that he will complete his term. What he has demonstrated with clarity is his support for the exploitation of precious metals, which he justifies through the need for money and jobs.

Guatemala is in the midst of a terrible crisis of impunity and lack of confidence in public institutions, civil servants, and the justice system; and hundreds of licences to explore and extract precious metals remain on the books. At the same the time, the 1997 mining law remains in force, granting mining companies free water, failing to protect Guatemalans' rights, ordering that the companies' 1 per cent taxes be paid on the basis of the legally certified statement of earnings of the mining company's representative. Daniel Kappes, president of Kappes,

Cassiday, and Associates (KCA), the company that, through its Guatemalan subsidiary Exmingua, exploits gold and silver at the La Puya mine, just outside the capital, recently published a full-page statement in *Prensa Libre* in which he assures us,

> In Guatemala there are powerful non-governmental organizations that consider socialism, terrorism, and anarchy to be the best forms of government for a society. It's been proved that radical NGOs are a profitable business. The people who run them earn big salaries, produce nothing, and don't pay taxes.... They are the true enemies of this beautiful country. We believe that, within the framework of the state of law, their negative activities should be curtailed.... The NGOs and their violent acts make it difficult to fulfil the government's constitutional mandate (Article 25 of the Constitution declares the rational, technological exploitation of hydrocarbons, minerals, and other non-renewable resources to be of public utility; the state will establish and provide appropriate conditions for their exploration, exploitation, and commercialization), often acting on the margins of the law. Guatemala has a great future and I'm sure that, aside from the unrest caused by these organizations, Guatemalans are, and will be, proud of their nation.

This mining company Kappes, Cassiday, and Associates (KCA) spends US$5,000 on a publication riddled with falsehoods and disobeys judicial injunctions that order a halt to its operations, because it knows it has the support of the government, which assigns police and soldiers to look after the company's interests. These days news about social conflicts over mining barely appear in the mass media. I myself can no longer continue making public denunciations of how precious metal mining is affecting the communities on which it is imposed, for in November 2015 I received a brief surprise note from the secretary of the administrative department of *Prensa Libre* in which she indicated to me that the newspaper had decided to do without my columns.

Adapted, from columns published in the newspaper Prensa Libre, *and translated from the Spanish, by Stephen Henighan*

5 Impunity in Guatemala: A Never-Ending Battle

HELEN MACK CHANG

The New Battles against Impunity and Corruption

Guatemala is experiencing a profound political crisis, which is spilling over into society and will affect its economy. This became glaringly apparent on 16 April 2015, when the UN's CICIG revealed what may be the greatest corruption scandal seen in government, extending into the highest spheres of this administration.

The private secretary of Vice President Roxana Baldetti was directly accused of leading an extensive criminal fraud ring within Guatemala's tax-collection agency, the Superintendencia de Administración Tributaria (SAT; Tax Administration Supervisory Office). He has since resigned. The fallout: the private secretary is on the run, two superintendents and some mid-level bureaucrats and SAT workers have been named in the proceedings, and the vice president has resigned. Since the beginning of 2014, some of the most highly placed politicians in the country have been manoeuvring to ensure that CICIG's mandate is not renewed and to halt the streamlining of the ongoing battle against impunity and other scourges that affect the state and society in general: corruption, influence peddling, drug trafficking, and organized crime. CICIG's mandate was due to end in September 2015, and the president of the republic, Otto Pérez Molina, led a campaign to prevent its renewal or castrate its operating capacity so that it had no real way to affect structural impunity or to fight against criminal organizations. Such organizations have co-opted the state and are seeking to reconfigure it to benefit corrupt, criminal influences. Calls of "No to CICIG!" began to be heard in the media, and it appeared that the campaign was succeeding.

As has become apparent, these efforts by the current government were intended to pave the way, through the removal of CICIG, for public institutions to be completely co-opted by hidden power structures. These structures are known as Cuerpos Ilegales y Aparatos Clandestinos de Seguridad (illegal groups and clandestine security organizations). Born out of the military intelligence apparatus during Guatemala's civil war, they have mutated to become organized crime networks, with participants from every sector of society, both public and private.

Their activities, which have now gone well beyond contraband and illegal trafficking, have consequences within political power structures and societal organization, acting on and destructively influencing many aspects of life in the country. They promote illegal actions by corrupt businesses exploiting natural resources, whose inappropriate use is seriously harming the environment. For example, influence peddling made it possible for a company inappropriately and inefficiently running an African palm plantation to cause an environmental disaster by spilling toxic substances into the La Pasión River in Sayaxché in the northern Department of Petén, along the border with Mexico. The chemical and organic contamination in the river caused a massive die-off of fish and other aquatic animals in the ecosystem. This type of pollution is considered to be the most serious environmental problem in the country, and its main impact is its negative effects on the ability of local populations, who depend on traditional fishing, to earn a living.

Another significant example of the anomalous and illegal operations of these groups is an illegal jade-mining operation run by people who do not have a mining permit from the country's Ministerio de Minas e Energía (MEM; Ministry of Energy and Mines). The paradoxical side of this case is the fact that Liu Jung Chih and/or Luis Lio, who are Asian, and Gustavo Salvatierra, who is Guatemalan, were extracting large amounts of jade from the municipality of Usumatlán, Zacapa, without a permit, but were nevertheless being protected by the Policía Nacional Civil (National Civil Police). On another note, because illegal groups have infiltrated the judiciary, co-opting corrupt judges, community leaders who are fighting mining and hydroelectric companies for control over their own lives and lands are now being treated like delinquents before the courts. Their protests and allegations are criminalized to the benefit of these companies, which have been infiltrated by the same groups. This is what happened to community leaders Adalberto Villatoro, Francisco Juan, and Arturo Pablo from Santa Cruz Barrillas,

Huehuetenango, who have been arbitrarily accused of abduction or kidnapping, threats, inciting criminal acts, and holding illegal meetings and demonstrations, all to crush their social movement with fear in order to favour hydroelectric company Hidro Santa Cruz (the criminalization of protest is discussed by Magalí Rey Rosa in chapter 4).

In mid-April 2015, scandal exploded when it became public that a criminal ring was being run from within the office of the president, presumably with the knowledge and participation of the president and vice president of the republic. Members of the ring from both the public and private sector were revealed to be responsible for a multi-million-dollar tax fraud, with all of the political, ethical, legal, financial, and institutional repercussions that this entailed. However, it is naive to believe that such a significant event would end there. On 25 April, for the first time in more than thirty years, the urban middle class poured into the streets and took over the Plaza de la Constitución in a peaceful protest. They were rejecting the government and corruption, and indirectly, the political system itself, and they opened a profound and probably deadly wound in the oligarchic-military regime that has been in power since 1954, protected by the political system and its players. As the trial moved forward, it led to a cascading series of events that shook the executive body, beginning with the resignation of the vice president of the republic. Followed as it was by the resignation of ministers, secretaries, and bureaucrats, as well as by the appointment of a new conservative vice president to the Congress of the Republic, this left the president completely alone and isolated from the main, actual sources of power, almost unable to fulfil his role. The only coherent political outcome is resignation, which he continues to resist.[1]

The pressure from politicians and regular citizens was unrelenting, and it soon turned towards the judiciary, which is on the defensive and practically unable to protect the criminal organizations by which it has been infiltrated. Judges are the subject of pretrial investigations, some members of these criminal rings have been exposed, and a Supreme Court justice is on leave, attempting to evade responsibility and refusing to face justice regarding the fact that many of her family members are implicated in these crimes. Next up was the legislative body, which is being pressured to approve legislation to enact urgent reforms to settle and rebuild the country's political system, in the midst of more requests for pretrial investigations against key members of Congress. They are accused of serious crimes such as unjust enrichment, influence peddling, abuse of authority, consorting with criminals, and fraud.

This breakdown, and the consequent changes, represents a unique opportunity to fight for participatory democracy. However, there is also a risk that authoritarian structures will double down or that stopgap measures will be adopted that could lead to an even greater crisis in the near future. To achieve democracy, we are participating, through institutions and personally, in any of the political spheres we can access in order to promote the most reasonable positions that could effect the transformations that the country needs. Not much has been defined yet, but there is a deep-seated process underway that is breaking down the structures that enable the impunity and corruption that run deep in the political and economic systems. As a consequence, those who benefit from this corruption are attempting to salvage the situation and their positions wherever they can. These structures have created the opportunity for this profound crisis. Such a state of affairs has led to tolerance, complicity, and cover-ups for the crimes of today and the past. Still it is possible that opportunities to make changes can be found.

We are therefore living in a failed political regime, in a culture where impunity and corruption are slowly gaining social consensus that makes it possible to legalize the illegal and obtain social and political power for a state that has been completely co-opted in service of organized crime and has stepped away from seeking the public good, the fundamental goal of democracy.

When people try to classify the impunity that arises from within the state itself, they say that we have legal and factual impunity, through the justice system and the impunity that is fomented by the people when they fail to seek justice after being the victims of crimes or having witnessed them. This is an arbitrary classification, but in my opinion it is useful, if only because it illustrates factors in the culture of impunity that affects us. There has been no political will to invest in the institutions that are in charge of security and justice, to fight a sustained battle against corruption, or to allow the normal development of trials seeking to clarify past events. Impunity in Guatemala is "un problema estructural derivado de la situación económica, social y política" (a structural problem arising from the economic, social, and political situation) (Rodríguez Barillas 1996, 3) which is expressed in structures that allow the resources and wealth of the country to be concentrated in few hands, with industries responsible for deep and far-reaching exploitation of the people. This leads to a socio-economic landscape that creates poverty and a lack of goods and services, and the majority of people live in a state of illness and ignorance. This is structural violence, and impunity is necessary if it is to prevail, because it is a system that can be maintained only through randomness, fear, and violence. This is our context as we continue to fight impunity and demand democracy and the rule of law, seeking a better life together for all members of society, with justice and equality.

Painful Beginnings

Twenty-five years ago, my life took an unexpected reversal when I found that my sister, Myrna Mack, had been stabbed to death by a special intelligence commando from the EMP. It was September 1990, during the last year of the first civil government after more than three decades of military rule. Levels of impunity were already high, and during that period there were both attempted and successful assassinations targeting journalists, politicians, and religious leaders. These cases became mired in legal technicalities, malicious legal action, and failure to investigate, and were never settled in court. The blanket of impunity had settled over many of these events, and I didn't want my sister's death to be one of them.

In the midst of pain and uncertainty, I began to exercise my right to petition the institutions in charge of investigating crimes and dispensing justice. I was not prepared to allow my sister's name, life, and work to be lost in that labyrinth of silence. I knocked on every door I could in my quest for justice. That quest turned into the expression of the battle against impunity it gained the support of others who had lost loved ones or had been the victims of human rights violations. I did this in a country where this still is a dangerous act, and where those in power even consider it subversive.

The case to investigate the material and intellectual authors of the murder of Myrna, and see them sentenced in court, went on for fourteen years. During this time, we faced threats, intimidation, and exile. The main police officer investigating the case, José Miguel Mérida Escobar, was murdered. Other events prevented the trial from moving forward, especially through the use and abuse of *amparo* appeals (constitutional appeals based on the violation of fundamental rights), which were filed to cause confusion and exhaustion in an attempt to have these crimes go unpunished. These appeals are legal, but they are ethically unacceptable, because their only aim is to delay the workings of due process.

Although the case made its way through the judicial process to determine who the perpetrators and instigators of Myrna's murder were, and it won in every court where our legislation allowed it to be heard, it remains an instance of partial impunity. One of the material authors, Noel de Jesús Beteta Álvarez, a special sergeant-major of the security section of the EMP, was convicted. However, one intellectual authors Colonel Juan Valencia Osorio, who was sentenced to thirty years in prison for giving the order to kill my sister, is now a fugitive. He escaped with the help of his colleagues and with the implicit tolerance of army authorities.

My aim here is not to tell my entire life story. My case is one among thousands. One must simply recall the long and terrible history of human rights violations in Guatemala. Sadly, impunity continues to gain ground.

The Search for Justice

Acts committed during the internal armed conflict are before the courts – acts involving security forces, police, officers, and troops of the Guatemalan Army, and paramilitary groups that are organized and directed by the army, like the Patrullas de Autodefensa Civil (Civil Defence Patrols), quasi-military groups of local citizens enlisted during the civil war to patrol their neighbourhoods to watch out for guerrilla activity. However, there are voices that deny the atrocities that were committed, especially against Indigenous people, who were displaced from their place of origin, or "disappeared." Family members of the victims of this repression have tirelessly sought to know why these acts occurred. As long as access to the truth and to justice is denied, taking away from the children and grandchildren of the victims the opportunity to acknowledge and honour them, and when those who are responsible are not brought to justice, impunity is fomented, the rule of law is dealt a severe blow, and the proper administration of justice is threatened.

The genocide trial of former president and retired General Efraín Ríos Mont, which took place through 2013, provided proof of how polarized our society is, and the same accusations and arguments once again reared their heads to justify the crimes committed when the country was at war. The difference was that the accused and the accusers were facing a court, giving their version of events. For the first time the victims formally expressed before the judicial institutions of the Guatemalan state the pain and suffering that they had endured, telling the tale of their tragedies. The accused pleaded innocent or claimed not to have been present where massacres had taken place. The court listened to the testimony and had access to documentary evidence and expert witness reports. It handed down a ruling in the midst of accusations that it was not impartial, accusations made by the representatives of the de facto power groups that need and bring about the structural impunity of the state. The magnitude of a guilty decision against a former head of state, a member of the self-designated and untouchable military, defender of the most conservative and authoritarian economic interests in the country, showing the perverse exercise of power by the Guatemalan State, was not well received by the financial-business block, and without hesitating, it ordered that the Corte de Constitucionalidad (Constitutional Court), the most important court of justice in the land, to overturn the decision. Ten days later, the sentence was overturned, and impunity was strengthened yet again.

The team of lawyers defending the army halted the trial. It had been set to begin in January 2015, but was prevented from doing so. The Instituto Nacional de Ciencias Forenses (INACIF; National Institute for Forensic Science) ruled that General Ríos Mont's health was too poor for him to withstand another trial because he was so elderly. According to the report, "He is not fully able to exercise his mental capacities, he is not able to properly understand an accusation made against him, he is not able to understand the parts of a trial and legal proceedings, and he is not capable of participating in his own defence. As a consequence, he is not competent to attend or pay attention to court hearings."[2]

Other cases of human rights violations, such as the massacre of Las Dos Erres, the massacre of Xamán, or the case of the Diario Militar, had to be heard at the IACHR in order to seek the much-longed-for justice before that international court, because domestic courts were unable to deal with the victims who had survived these events. In these cases, the Guatemalan state was condemned for denying justice and for participating in these crimes.

Today in Guatemala there is no separation of powers, and judges are arbitrarily imposed on the country by power groups to guarantee their impunity. These groups have no compunction in interfering in the selection of authorities of the institutions in charge of dispensing justice. They don't mind flouting the laws governing nominations, or showing the political negotiations that take place between the Congress representatives from different groups. They care little about honour, capability, or track record regarding integrity as the minimum criteria for those who would sit as judges. They force their choices upon the people, with shady alliances, *compadrazgo* (an entrenched system of

reciprocal granting of favours based on personal relationships that creates informal political patronage), and influence peddling, all to make
it easier to commit future crimes.

To all this we must add that studies done by the Ministerio Público
(Public Prosecutor's Office) show that, when impunity is at its worst,
98 per cent of crimes go unpunished. However, work to report these
issues and support ongoing action to keep the pressure on the MP, with
the help of CICIG, especially during the tenure of Amílcar Velásquez
Zárate and Dr Claudia Paz y Paz, made it possible to work together to
manage cases more efficiently during the investigation.

It has been said that justice in Guatemala is the Cinderella of the state
because those whose work is dispensing justice receive so little support.
But now the mafia, which has become embedded within the state, has
made justice into a prostitute, one who dances at a ball and costs millions
of dollars, who can guarantee impunity to the highest bidder. Even so, in
the justice system we have judges and prosecutors who are and committed to applying the law, and who have done good work in spite of many
challenges and a lack of resources, while risking their own lives. These
people give justice its true value, which it should have for everyone.

How Far Does Impunity Go?

Impunity has transcended civil society and threatens to install corruption as a widespread and accepted social practice. Our political system
has become grotesquely corrupt, leading the way to a profound sense
of mistrust in all citizens towards "the democratic process," which is
limited to our going to the polling booths and voting once every four
years to elect new authorities. This year, 2015, is an election year in Guatemala, and, as in previous elections, candidates reaffirm their "commitment" to combatting impunity. However, such offers will remain
nothing more than campaign promises, which will go unfulfilled in a
shameless display of evil. There are no plans for governing that are
rooted in the reality of the country, and candidates are vote-shopping
by cherry-picking benefits for narrow groups of voters.

Another example of the impunity with which politicians behave is
the failure to comply with the Ley Electoral y de Partidos Políticos (Electoral and Political Party Law), or with the Tribunal Supremo Electoral
(TSE; Supreme Electoral Court), the highest authority, which paradoxically, in the face of so many brazen acts, is no longer supreme. Both
members of Congress and public employees campaign permanently

flouting the terms and prohibitions set forth in the law. Their shamelessness is such that Congress has become paralysed, unable to do its job, and is now central to election and industry deal-making. As we have seen in the media, "On the hustings, mortal enemies; in Congress, political allies."

In the Congress of the Republic the national list of jobs to be done in the different regions of the country is divided among the political parties, most for the benefit of real or fictitious NGOs that are linked to members of Congress and their family members and/or friends. This is how they get around the Ley de Compras y Contrataciones (Procurement and Contracting Law), and they have no qualms in doing so.

The lack of security, along with impunity for those who break the law, translates into violence that has continued to increase. Extortions against large and small companies have become a scourge. The most worrisome result is that anyone who refuses to pay or is late in paying is threatened from within the prisons, where collections on monies extorted are organized and death threats are issued. These criminal structures act with the collusion of the authorities, taking advantage of an unmonitored and inept penitentiary system. Those who control the prisons with complete impunity are the prisoners themselves, many of whom are middle managers or top executives in these criminal rings. Guatemala's prisons have become the best universities for organized crime and terror.

Meanwhile, the Policía Nacional Civil (National Civil Police) does not even have gasoline for its vehicles. I highlighted the precarious conditions in which they work when I coordinated the first part of the Comisión de Reforma Policía (Police Reform Commission), and little has changedn. The police do not have adequate offices and stations, and the agents and officers do not receive proper training that would permit them to behave like professionals; even basic training is deficient. Police structures are vulnerable to being infiltrated by corruption and organized crime. This is especially true for organizations linked to drug trafficking. Combined taskforces were created with the participation of the PNC and military units, bringing to the fore the clear and constant control of the military. However, militarization reinforces a system of impunity and is a source of human rights violations. We are still dominated by a reactive and authoritarian system, intended to control the people, not a preventive system that allow the people to regain trust in security forces.

This battle against impunity, which seems never-ending, motivated us to create the Alianza Contra la Impunidad (Alliance against Impunity), through which, we have begun to study this phenomenon to enable stakeholder advocacy groups to dovetail as we seek to strengthen our justice system. Together with other organizations, we have also created the Movimiento Pro Justicia (MPJ; Pro-Justice Movement), which monitors selection of the highest authorities in the justice system. The MPJ has been working on a proposal for a Ley de Comisiones de Postulación (Law for Nominating Commissions) for the main courts in the country, which will define the criteria for establishing these commissions and the procedures to appoint candidates for judges. The last time this occurred was in 2014, and the appointment devolved into a murky process burdened by accusations of shady deals between the governing party and the main opposition party. The appointment became politicized along party lines, practically destroying the validity of a process-based and methodical candidate selection procedure, whic had met the minimum requirements for selecting authorities who were at least mostly independent and could be asked to battle impunity and to properly administer justice. A clear example a complaint filed by Judge Claudia Escobar. Together with other judges, she reported irregularities in the appointment process and resigned from her position. It became evident that a top executive of the governing party, Congress member Gudy Rivera, had interfered in the process. In a meeting set up by a judicial adviser, he suggested that Judge Escobar rule in favour of an appeal to benefit Vice President Roxana Baldetti. In exchange, Escobar was to be re-elected to her position as judge. This triggered a political scandal, revealing the behind-the-scenes machinations that are still taking place at the highest level.

With the same goal, having seen how far impunity could reach, we lobbied for the creation of an International Commission, in compliance with the Acuerdo Global sobre Derechos Humanos (Comprehensive Agreement on Human Rights) signed in 1994 as part of the process for the Peace Accords. It would be able to investigate the illegal organizations and clandestine security apparatuses named in the CIACS Agreement. The discussions and context within which the proposal developed arose from what we know today as the CICIG, which was organized at Guatemala's request by the United Nations at the end of 2006, making it an international mission within the purview of the organization's actions on behalf of human rights.

An Epilogue

The impunity of the past is the impunity of the present, and the most lamentable part has extended its influence into every corner of society. The Peace Accords were signed in 1996, to become the goals for the country in dealing with the problemse discussed during the peace process. They now seem far away. Almost twenty years after signing, the topics related to justice that were raised at the time are still being discussed, in the midst of the institutional crisis. Today, this crisis could represent an opportunity for Guatemala to rise up once again. It's a time for people to say, "Enough!" Enough impunity, corruption, self-interested and indolent politicians, and mafias infiltrating public government. All of the evils of society that affect us, and against which we have fought and will continue to fight, come together in impunity. Nobody is above the law, and the rule of law must prevail.

Through these new battles, the people have lost their fear. Through social networks, young people have found agency, and good proposals will come from their enthusiasm and new ideas. People of all ages who reside in rural areas and who come from all industries have joined the spontaneous protests of the urban middle classg since 25 April 2015. By reclaiming rights and expressing dissatisfaction, a baton can be passed from one generation to the next. People are rejecting an entire rotten system, which requires a purge of other state organizations and the political class.

It has not been easy to get through this time in the midst of a climate that deteriorates day by day, and that many sectors of the population are beginning to accept as normal. However, there are points of resistance where people refuse to give up in the face of this near-tragic situation. After twenty-five years of a battle against impunity, it could seem as though frustration or exhaustion had taken away our will to continue. Sometimes we feel that there is little we can do, yet our belief is stronger that it is possible to create a fair and liveable society. We are unwaveringly commited to the younger generations so that they will have access to life within a different social reality. They will become conscious of the role that they must play to build the fair and democratic society that we all deserve.

Translated from the Spanish by Lisa Maldonado

NOTES

1 This chapter was written prior to the resignation of President Otto Pérez
  Molina on 2 September 2015. Accounts of the resignation appear in chapter
  1 by Candace Johnson and chapter 9 by W. George Lovell.
2 In chapter 1, Candace Johnson updates the Ríos Montt trials.

REFERENCES

Rodríguez Barillas, Alejandro. 1996. *El problema de la impunidad en Guatemala*.
  Ciudad de Guatemala: Avancso/ Fundación Myrna Mack.

6 Politics, Institutions, and the Prospects for Justice in Guatemala

CLAUDIA PAZ Y PAZ

Guatemala lived through a civil war that lasted more than three decades and ended in 1996 with the signing of the Peace Accords. It was a war in which serious and massive violations of human rights were committed. The Historical Clarification Commission (or CEH, also known as the Truth Commission), recorded more than 42,000 victims of extra-judicial executions, forced disappearance, and torture, violence, and rape, including men, women, and children. Of those who were able to be fully identified, 83 per cent of the victims were Mayan and 17 per cent were Ladinos. Combining these data with other studies examining political violence in Guatemala, the CEH has estimated that more than two hundred thousand people were murdered or "disappeared." The commission also revealed that between 1981 and 1983 genocide was perpetrated in regions of the country against the Mayan population. One challenge of the peace process and Peace Accords was to rebuild the justice system, changing it from one that had covered up the most serious human rights violations into one that could fulfil its essential function of protecting its citizens and guaranteeing access to justice for all.

Within approximately ten years, a number of justice institutions were created: the Office of the Public Prosecutor (MP), the new National Civil Police (PNC), the Instituto de la Defensa Publica Penal (Institute for Public Criminal Defence), magistrates' courts, and several trial courts in parts of the country where there had been none.

However, ten years after the signing of the Peace Accords, the promise of guaranteed access to justice for all was still far from having been met. On the one hand, levels of impunity remained extremely high, and most cases, especially the most violent ones, including homicides and many cases of violence against women, remained unpunished. At the same time, homicide rates continued to increase, as they did in neighbouring countries in Central America. At their highest point in 2009, rates reached levels of 46 murders per 100,000 inhabitants.

Two factors can illustrate the climate of violence that Guatemala experienced at that time: one was the murder of seven people who were incarcerated at the Pavón Prison, which was perpetrated by members of the National Police force; in fact, this event has already resulted in the then-director of the PNC being sentenced in Switzerland; Carlos Vielman, who was the minister of the interior at the time, is awaiting trial in Spain. These were crimes committed by highly placed state officials.

The second event is the murder in February 2007 of three members of the Central American Parliament and their driver. They were murdered by members of the police force who themselves were murdered in one of Guatemala's maximum-security prisons.

These events illustrate the climate that reigned at the time in Guatemala, where there was a lack of trust and belief in justice officials and security forces, who were clearly involved in criminal actions. In the face of so much violence and insecurity, the people of Guatemala began publicly clamouring for action against these and other crimes – lynchings, murders, execution-type assassinations – even action that bordered on illegality. We also became witnesses to infiltration by organized crime into the country's security forces and justice-related structures.

As a result, Guatemala sought help from the United Nations, and in 2007 the CICIG was created and began to operate in the country. Thanks to its presence, significant changes were made to the highest authorities of the justice system. New officials were appointed to the Supreme Court of Justice, the Ministry of the Interior, and the MP, with the appointment of a new public prosecutor, and the previous process by which public prosecutors had been chosen was annulled, and under the new regime I was elected.

This made it possible to create unprecedented synergy in the country, allowing for cooperation between the police, public prosecutors, and judges. There were many successful prosecutions in organized crime and drug trafficking, and investigations against gang members who were involved in extortion, as well as some cases of violence against women that were solved. There was progress in cases that had been winding their way through the courts for years, especially those involving human rights violations.

Throughout these years, the victims and survivors of the war kept up their clamour for justice. The first lawsuits involving genocide and crimes against humanity were brought before the courts in 2000 and 2001.

Along with the ceaseless calls for justice of the victims, several sentences were handed down in the Corte nacional (National Court), clearing the way for judgments in cases of severe violations to human rights. One important ruling of the Constitutional Court (CC) declared that military documents were not secret and should be made available to investigators; along with the discovery of the archives of the National Police, this provided documentary evidence for a number of different cases.

The Inter-American Court of Justice ordered that obstacles in fact or in law that had impeded these crimes from being judged must be removed.

In addition, work had been done for more than twenty years to exhume graves and record the scientific evidence that would later be used in the genocide case. More than 8,000 bodies have been exhumed in Guatemala. These exhumations in hundreds of cases would constituted forensic evidence of the brutal genocide.

In January 2013 a Guatemalan judge took the daring step of ordering that a criminal trial should begin for genocide and crimes against humanity against former president Efraín Rios Montt. And it began in the following March.

The case went to court supported by compelling evidence. More than forty expert witnesses and more than one hundred eye witnesses testified, and forensic evidence was submitted for consideration.

Perhaps the most emotional part of the trial was the moment when ten survivors of sexual violence gave their testimony.

It had been requested ahead of time that they be allowed to give their depositions behind closed doors to avoid re-victimizing them. In some cases not even their families and communities knew that they had been sexually assaulted. The biggest question, which was hotly debated, was how to introduce this important evidence without victimizing the women and violating them again. I clearly remember the words of one colleague, who said, "I don't know what you're talking about. These women are tougher than we are. Let's ask them and see what they say. Let them decide if they want to give testimony."

The women testifying were indeed strong, and they gave their testimony. Something changed after that day. The media blackout on the case was broken. For all Guatemalan citizens who saw these women and heard their stories, it was impossible not to feel empathy with their pain after witnessing their bravery.

They and the other witnesses to the trial, who had travelled from the most remote mountain regions of Guatemala, came to the highest justice institution in the country, the Chamber of the Supreme Court, where the hearing was taking place. There, in their own language, they faced Ríos Montt, who had ordered that these human rights violations be committed, faced the man who had commanded the army and who had once had all of the powers of the state; and they told him, "You did this to us. This is wrong. We are people, not animals."

He was forced to listen to them before the court.

Justice has an amazing power to equalize. The trial finally made it possible, for once, for the perpetrator and the victims to meet on equal footing and for all to experience the equalization brought by justice.

Through the trial the silence was broken around many human rights violations, especially around those involving sexual violence.

The trial had huge healing power for the victims. The blame that they might have felt after having experienced these abuses to leave the bodies and lives of these women came to rest with those who were truly responsible – the perpetrators.

On 10 May 2013, "Tribunal A" for High-Risk Cases in Guatemala City found General Efrain Rios Montt guilty of genocide and crimes against humanity.

A few days later, however, the sentence was annulled by the Constitutional Court, for reasons that are incomprehensible. The defence team made no attempt to create a solid case. Instead, it discredited the victims, the human rights organizations, and the officials, including myself, who had supported the process. Their strategy was to have a lawyer who, on the first day of the trial, stood before the court and stated that he was an enemy of two of the members of the court and a friend of the third; he requested that the court declare it could no longer hear the case. The response of the court was of course that he was wrong, and that if the lawyer was a friend or enemy of any of its members, the lawyer should be the person who recused himself. This was the basis on which the CC annulled the trial.

In Guatemala, as in many countries in the world, the Constitutional Court has only extraordinary jurisdiction: it should hear cases only when all other actions and appeals have failed before other courts. In this case, it became involved in the deliberations directly with great haste and annulled the trial.

Having reached this point, after the enormous efforts of the victims, lawyers, prosecutors, court, and supporters in the international community, after the sentence and its annulment, this all led to one question: Was the battle of the victims worth it? The bravery of the judges and the prosecutors, the solidarity shown throughout the entire trial?

One legacy of the trial was an effort to deny the truth, for Guatemalan society to avoid facing that there had been genocide in the country and that a sentence said so.

The most scurrilous event was a decree by the Congress of the Republic saying, "There was no genocide." It's as if they had handed down a decree saying, "By decree there is no hunger in the country. By decree there is no violence." This decree was an affront to the republican concept of separation of powers, because these matters were before a court and there was a trial pending, and a Parliament can say, "This is a crime," but it cannot say whether something happened, or whether someone is guilty (Catherine Nolin, in chapter 2, describes reactions to this decree among victims' relatives).

Nevertheless, paradoxically, in the face of these efforts, you can't fool people forever, and there was a poll, such as polls before elections, in which they asked Guatemalans whether they believed that genocide had occurred. Over 80 per cent of Guatemalans did have an opinion. About 36 per cent of Guatemalans believed that there had been no genocide, and 47 per cent believed that there had been genocide.

Another consequence for the bureaucrats in the trial is that we were the victims of a campaign to discredit us. Almost every Sunday, we opeed up the newspaper and saw pictures of our own faces and the faces of our families. People said that we were subversives, that we were guerrillas, that this was not justice, but vengeance. The judge who presided over the trial and handed down the sentence was accused and attacked by the national law society unfairly.

However, in the midst of these campaigns to discredit us and paint us as criminals, the huge gain that came out of the trial and the sentence was a strengthening of the rule of law. It was proven that it is possible to bring a former head of state to trial in a domestic court, with a case that was very strong and solid. This was possible because there were honest officials, judges, and lawyers who would not be cowed and who fulfilled their duties. This was possible because of the huge effort and bravery of the victims, who never gave up on their fight for justice in Guatemala.

Transcribed and translated from the Spanish
by Lisa Maldonado

PART THREE

Cultural Responses to Injustice

7 Scars That Run Deep: Performing Violence and Memory in the Work of Regina José Galindo and Rosa Chávez

RITA M. PALACIOS

Performance is a two-edge sword, as powerful in contesting power as in maintaining it.

(Diana Taylor 2014)

El arte no puede hacer milagros, no puede cambiar una sociedad tan conflictiva como Guatemala pero sí es indispensable la expresión en un país como Guatemala.

(Art cannot perform miracles, it cannot change a conflict-ridden society like Guatemala's, but it is a must in a country like Guatemala.)[1]

(Regina Galindo 2014)

In May 2009 Rodrigo Rosenberg issued a posthumous plea for justice in an eighteen-minute video testimonial. In it he pointed squarely at Guatemala's president, Álvaro Colom, for what he alleged was his involvement in a conspiracy that resulted in the murder of his lover, Marjorie Musa, and her father, Khalil Musa, and ultimately of himself. The video made waves; the young lawyer ensured that 150 copies of it were distributed, and the video was shown on television and online immediately after his funeral (Calderón Bentin 2014, 364). Prompted by international pressure and widespread outrage, the president quickly denied any wrongdoing. Six months later, the CICIG would report that Rosenberg's claims were a complicated ruse crafted by a star-crossed lover who looked for justice in the only place he knew he could find it: the public sphere.[2] The public outrage originated in part from the fact that not even a well-known, well-liked rich Ladino,[3] with ties to the government and the business community, was safe from the reach of the state and its extrajudicial arm.[4] At a glance, it is easy to conclude that Rosenberg's approach to seeking justice was a direct response to Guatemala's atrocious human rights record and well-known history of impunity.[5] His decision to *perform* an injustice to arrive at a form of justice, even one attained through a court of public opinion, is of note as it shows that his acts were, in many ways, the *requisite* response.

On paper, the Guatemalan state promotes a democratic ideology and a regulation of public and private life through the just enactment of legislative, executive, and judicial powers. However, in practice, the state has been an instrument at the disposal of the few to the detriment of the many: with the legacy of a thirty-six-year civil war that was heavily subsidized by the U.S. government, supported by the oligarchy, and carried out by the military, the state sustains itself through series of acts that contribute to the illusion of the rule of law.[6] These acts, however, have worn through the fabric of their own artifice and their fiction is ever so present, resulting in a severe lack of confidence in the government and a general feeling of helplessness. As a scholar and practitioner of the law and a member of the upper echelons of Guatemalan society, Rosenberg understood the shortcomings of Guatemala's justice system and the state's parallel structures of power. It was precisely this intimate knowledge that shaped his performance, which was not as unexpected as we would like to think: he planned and carried out his own murder, taking a page from Guatemala's very own playbook, a playbook that includes violence, intimidation, conspiracy, but most importantly, fiction.[7] This is what makes Rosenberg's case unique: a fiction was created and sustained not by the state but by a private citizen. More importantly, this fiction managed to bring to light the state's complicity in a flawed justice system and ongoing impunity.

Rosenberg's answer was to resist the effects of the all-too-common practice of impunity that relies on acts of erasure for its continuous and lasting articulation. He turned to the camera, gave his testimony, and reached tens of thousands of Guatemalans who saw themselves in him: he created a performance that prompted his audience to tap into the feelings of helplessness and pain that had enveloped him, and, with their help, the performance became a collective denunciation of Guatemala's failed justice system.[8] Rosenberg's acts were, in part, a response to the systemic mishandling of knowledge of the Musa murders, but more importantly, they were a direct affront to the state, playing on its terms and using its language, demonstrating that the heavily mediated and theatrical stage on which politics is done in Guatemala needs to be met with a response that resists, contests, and makes evident its artifices.

My discussion of Rosenberg's decision to stage his own murder and sidestep ineffective legal channels brings up the interplay between memory and performance, and archive and repertoire that is important in addressing issues of justice and violence in Guatemala today. It also helps set the stage for a study of the work by performance artist Regina José Galindo (b. 1974), and poet Rosa Chávez (b. 1980), whose work resists official and collective oblivion by simultaneously staging the unsayable and lending a body to the memory and the trauma of post-conflict Guatemala. Moreover, the Rosenberg case lays bare the complicated script from which Guatemalans have been obliged to read since the signing of the Peace Accords in 1996. Under the guise of democracy, neoliberal policies have been implemented, genocide denied, and corruption further perpetuated. In some ways, Guatemala's civil war has been handled as an open-and-shut case, despite the obvious need for actual healing, reconciliation, and repair. The violence that is a part of everyday life has taken on new forms, and no single specific culprit or group is responsible, as in times past. What is perhaps most troubling is that there seems to be a selective amnesia among certain groups in the country.

The work done by international and national bodies, grassroots groups, and activists, to name a few, has ensured that recent history is not forgotten, and they continue to expose corruption and fight for justice. And though their work in preserving the past and initiating a process of healing is of paramount importance, the steps that they undertake, collecting first-hand accounts, as did the CEH,[9] for example, can never capture the intensity and impact of the violence and trauma. These formal processes categorize and archive the individual and collective memories that can be *recorded* but overlook what can only be *experienced*. In other words, what cannot be wholly addressed in formal legal proceedings, such as the unsayable aspects of pain, fear, and loss, can be lost: the memories and feelings of the witness-survivor, for better or worse, become part of a larger narrative that defines truth and measures violence, relegating the unsayable to a distant memory, an uncomfortable silence, or a willed amnesia. And so, while reports, trials, and sentencing may bring justice, they are not enough to address the lasting effects of violence.[10]

Two formal inquiries were carried out shortly after the Peace Accords were signed, one by the Catholic Church's REHMI project and the other by the CEH, and they issued a series of recommendations to new post-war governments but were quickly repudiated or ignored.[11] In 2012, Amnesty International declared that "despite the welcome progress in [the Dos Erres massacre] and a handful of other cases, much remains to be done to achieve justice, including advancing cases against those with command responsibility for serious human rights violations" (2012, 5). REHMI's and CEH's findings have played a role of utmost importance in bringing to justice those who committed heinous crimes during the war period, and many organizations have used these findings to better support victims, survivors, and their families. However, these reports have not been followed up by sufficient government action or supplied with the necessary structural support, which, as a result, has redirected them to the realm of forgetting by being "shelved" much too soon.[12]

What is needed are mechanisms that prevent the unsayable from being obliterated, tap into collective memory, and recall certain aspects of the unspoken experience. The work of Regina Galindo and Rosa Chávez functions as such a mechanism, for it deals with the violence and the attempts, intentional or otherwise, to erase the unspeakable aspects of the war. Their work also provides a contrast to the state's theatrical enactment of the rule of law, which is shown to rely on artifice for its own legitimization, and violence for its enforcement. Using her body as more than a canvas, sometimes presenting it as evidence, a history book, a laboratory, or a testimonio,[13] Galindo provides her public with a site on which pain, suffering, and visible and invisible wounds can be remembered and safely experienced. Much in the same manner, the performative aspects of Chávez's work reaffirm the resilience of Indigenous people, acknowledge the effects of the war, and contest the continued colonial practices that the discourse of the neoliberal state legitimizes. Their collaborations, "Hermana" (Sister) and "Inmonumental" (Unmonumental), are significant because they further challenge the sometimes homogenizing effects of a human rights discourse and state-sponsored silence. Specifically, these works add an important dimension to the discussion on the issues of war and violence, memory and trauma: the often conflicted relationship between Ladinos and Indigenous peoples that has shaped Guatemala for more than five hundred years.

Performing Memory and Trauma in Guatemala

In our society, agents or former agents of the state have woven a secret, behind-the-scenes network dedicated to obstructing justice. They have created a virtual alternative government that functions clandestinely with its own standardized and consistent modus operandi. In such a context, crimes are not clarified, and those responsible are not identified. Society finally forgets the cases and becomes resigned.

(Institute of Political, Economic and Social Studies, 2000)[14]

When I speak about performance and memory, I utilize Diana Taylor's concepts of archive and repertoire. According to Taylor, the archive is composed of "enduring materials" that include texts, maps, documents, and buildings, and is believed to be long-lasting and immutable, given that in Western society the written word is imbued with knowledge and power (2013, 19). The reports produced by the two inquiring bodies in post-war Guatemala fall under this category, and although their importance cannot be denied, all on their own they cannot act as a vehicle for healing, for the reasons outlined above. Instead, Taylor's repertoire of "embodied memory" provides an alternative: it exists in "embodied practice," it is ephemeral, and it "requires presence: people participate in the production and reproduction of knowledge by 'being there,' being a part of the transmission…. The repertoire both keeps and transforms choreographies of meaning" (20). What is more, in the repertoire, "performances function as vital acts of transfer, transmitting social knowledge, memory and a sense of identity" (2). And even when the archive attempts to capture the performance, Taylor reminds us, it can never successfully do so because their natures are dissimilar (20).

Interestingly, given the reach of digital technologies, performance has found new venues and, as a consequence, the functions of the archive and the repertoire have been altered. This is the case with some of Galindo's and Chávez's work: while it is performed, it is also recorded and transmitted online where it reaches other audiences at different times, in different contexts, thus challenging the requirement of "presence" of the repertoire that Taylor identifies. In fact, what is presented online is markedly different from what once took place; it may be packaged to communicate the notion of collaboration or it may capture only key moments that the artists choose to highlight. Whatever the case may be, what is interesting is that Galindo's work first gained prominence as a result of its online presence.[15] Her best-known performance became key in the denunciation of Efraín Ríos Montt's presidential bid shortly after ¿Quién puede borrar las huellas? (Who can erase the traces?) was performed, recorded, and circulated online: "[A] group of artists began the necessary work: spreading word of the performance and the message. A curator friend of mine, Rosina Cazali, sent out images of the performance alongside a text declaring Ríos Montt's candidacy unacceptable.... After they were published online, the images of the performance were then published in newspapers that reached various groups" (Goldman 2006).

Galindo further underscores the difference between the performance itself and the record of it because "to present it on video is simply to show a document. In this case [i.e., *¿Quién puede borrar las huellas?*], whoever sees this document can come to know the history behind it" (Goldman 2006).

Nonetheless, it is important to note that, as Taylor later points out, with digital technologies come new forms of transmission of knowledge that challenge Western models and raise "new issues around presence, temporality, space, embodiment, sociability, and memory (usually associated with the repertoire) and those of copyright, authority, history, and preservation (linked to the archive)" (Taylor 2010, 3). Chávez, for her part, is an active participant in the arts community and has worked on a number of collaborative arts projects, some of which incorporate her poetry and are featured on video. In general, recorded readings of her poetry can be found online, and she keeps an active Facebook page and less active blog (santatirana.blogspot.ca) where she posts things of general interest as well as some of her verses. In this digital space, Chávez, like many other artists, complements her body of work and is able to expand her readership.

This new mode of transmission, with distinct methods and objectives, challenges the manners in which we interact with art and proposes new ways of looking at memory, as is the case with Galindo's and Chávez's work. For example, taking into account that many Guatemalans fled the country as a result of the war,[16] the use of digital technologies adds another layer of complexity, as it provides those abroad with the possibility of being "present" at the performance. The sense of the "here and now" that is intrinsic to the repertoire and is replicated in the "here of the web" allows the public abroad to access memory through the activation of the emotions associated with it:

> Place/thing/practice change online. Again, the three are deeply interconnected and altered in and through digital technologies. The spatiality of the archive as a "public building" gives way to the paradoxical ubiquity and seeming no-where-ness of the digital archive. The site-specific character of performance repertoires, that unfold in the here-and-now also

give way to the multi-sitedness of the web. We are all seemingly "here," live, now, online – no matter where the "here" might be. The "here" of the repertoire is immediate; the "here" of the archive is distant, but locatable; the here of the web is immediate and (only apparently) unlocatable. (Taylor 2010, 9)

On the web, viewers can access the performance time and time again, recreating its "here and newness," while the main function of the traditional archive, preserving knowledge in a tangible manner, is diminished because of the immediacy and abstractness of the digital archive. This is perhaps one of the greatest threats the web poses to the traditional archive: its immediate character relies on a "hereness" that is not spatially or temporarily defined, thus altering our relationship to the past and future. Paradoxically, the two artists tap into the "here of the web" to bring up the past and to shape a future – temporalities (and distances) that are extraneous to the perpetual present of globalization. While a further discussion of the online dissemination of Chávez's and Galindo's work is relevant, particularly as it pertains to a population affected by the war that accesses and deals with memory in many different ways, this study focuses on how their work negotiates issues of memory and justice, healing and reconciliation as a response to official narratives, "approved" histories, continued violence, and enduring trauma.

It is also important to point out a link between trauma and performance that goes beyond a simple exploration of memory. Specifically, there are certain key aspects that make one a vehicle for safely accessing and dealing with the other. Taylor outlines what she considers the shared characteristics of trauma and performance: performance allows survivors to cope with the trauma; trauma, like performance, is reiterative; they are both felt "affectively and viscerally" in the now; and they are "always in situ" (2013, 166–7). A special issue on trauma in *Performance Research* explores the importance of testifying, "bearing witness to one's own trauma" and being heard, which is an intrinsic part of theatrical performance where "the testimony is re-framed, the testifier re-personated" (Duggan and Wallis 2011, 7). One of the principal implications of this process is the involvement of the audience, who are urged to bear witness and, at the same time, are confronted with their own trauma and/or inaction.

In the case of Guatemala, where the process of memory is thwarted by the different actors who continuously shift the notions of truth and justice,[17] the notion of bearing witness inevitably brings up questions of ethics and morality. When audiences are faced with a performance in the streets of Guatemala City or Panajachel, not only is their memory activated, but they are also prompted to be present in the now, in the representation of the trauma. This results in their immediate involvement, whether they choose to observe, interact, or ignore. This last choice carries with it a moral and ethical weight in which guilt, responsibility, shame, and solidarity play a part. For Paul Antze and Michael Lambek, memory is a social performance that depends not on the individual but on a group of individuals to lend personhood intelligibility: "Memory is thus deeply implicated in concepts of personhood and accountability.... [P]ersons are corporate with regard to the estate of memory. Turning this around, it is memory and its tokens that provide the substantive grounds for claims to corporateness and continuity" (1996, xxv). Consequently, the refusal to participate in the process of memory is a denial of the experience of violence and the resulting trauma. It is also a denial of inclusion into a body politic in a post-war country.

Galindo's and Chávez's respective approaches require that the audience engage with their work; the type of engagement lies entirely in the audience's hands. This choice, in turn, points to the question of responsibility. From this we can surmise that the work of these two Guatemalan artists is anything but impartial: it prompts the viewer-reader to choose and, in so doing, identify her own role in the country's endemic violence, corruption, and racism. To this effect, I briefly turn to Hannah Arendt's notions of guilt and responsibility in her discussion of the participation of German people during the Nazi era. Arendt notes that in post-war Germany, those who were innocent often spoke of a "collective guilt," while those who directly participated may have claimed involvement but not guilt (1991, 276). In this case, what collective guilt reflects is responsibility, a recognition that involvement, or lack thereof, had very direct consequences upon others. This is where Arendt wishes to arrive: not to the incarceration of "guilty" millions, but a recognition of citizen responsibility in the years of the Nazi regime, because only in this manner can we become aware of what we, as humans beings, are capable of doing: "For the idea of humanity, when purged of all sentimentality, has the very serious consequence that in one form or another men must assume responsibility for all crimes committed by men and that all nations share the onus of evil committed by all others. Shame at being a human being is the purely individual and still non-political expression of this insight" (1991, 282).

In this vein, in one of Galindo's performances, *Punto Ciego (Blind Spot)*, a group of visually impaired participants enter a room where Galindo stands, naked, with no directive other than experiencing a work of contemporary visual art (Galindo 2012). The performance soon takes a turn for the unexpected as some of the participants begin to mistreat her, to the point of suggesting violence:

> Comenzó como una obra hipotética. Nos preguntamos: ¿cómo va a reaccionar un grupo de ciegos ante esta situación? Pero de pronto cobró un carácter muy político. Al principio fueron un poco tímidos al entrar. Pero después de que pasaron quince minutos y nadie les impuso límites, se armaron de valor, pues sabían que nadie que tuviera vista se encontraba en la sala. El grupo de personas ciegas se empoderó de la situación, y algunos se excedieron. Algunas de las personas comenzaron a manosearme, a agarrarme, a burlarse de mí. Me taparon la nariz y dijeron "tápenle la nariz, a ver si respira," luego dijeron "córtenla, a ver si sangra." Se hicieron los inocentes – como si no supieran que se trataba de una persona viva – y dijeron "no importa, es una muñeca nada más." La obra "Punto Ciego" nos demuestra que cualquiera puede tener la misma reacción que tuvieron algunos de los ciegos en aquel momento. Eso no es algo exclusivo de Guatemala. Puede pasar en cualquier parte del mundo. *Cualquiera es capaz de darle paso libre al lado oscuro de su carácter.* (Galindo 2012; my emphasis)

It began as a hypothetical work. We asked ourselves, How will a group of blind individuals respond to this situation? But suddenly, it took on a very political dimension. At the beginning, upon entering, they were a little shy. But after fifteen minutes passed and no one established any limits, they summoned their courage because they knew not one sighted person was in the room. The group of blind people took control of the situation and some of them overdid it. Some of them began touching me, groping me, making fun of me. They covered my nose and said, "Cover her nose, let's see if she breathes." Then they said, "Cut her, let's see if she bleeds." They played innocent – as if they didn't know it was a living person before them – and they said, "It doesn't matter, it's just a doll." The work *Blind Spot* shows us that anyone can react in the same manner that blind persons did at that moment. This is not exclusive to Guatemala. It can happen anywhere in the world. *Anyone is capable of giving way to their dark side.*

Overall, at the crux of Arendt's position on responsibility is moral shame, which is collective. A similar call to responsibility occurs in the performances of the two Guatemalan artists: their appeal is an invitation to share responsibility, or to at least define a position, which in turn helps create a community of individuals who may not necessarily experience personal guilt, but nevertheless are connected and can engage in a dialogue. This is an important aspect of their work, particularly since the state refuses to acknowledge responsibility and promote justice: the rejection of genocide is not only a refusal of any official responsibility, but also a denial of deep trauma and of the ethnic personhood of an entire people. While Guatemalans are left to their own devices to deal with the trauma and navigate the severely flawed system that they inherited, performance provides an inlet and an outlet, a place where what cannot be put into words can be accessed, brought out into the open, and safely explored.

Regina José Galindo: Performing Memory and Violence

The general population moves across public and private spheres with varying degrees of fear and trauma that stem from the recent past and the country's present reality. In many ways, violence is part of the everyday, as Guatemalans are bombarded by the media with explicit images of dead bodies and are often touched by violence themselves, seeing very little justice, much less support for coping with the resulting trauma. Some of the work that Regina José Galindo has presented in Guatemala takes place in public spaces, on the streets, where participants and audiences may not be immediately aware that a performance is occurring. Galindo is a well-respected performance artist who has gained international acclaim, and her work has been compared to that of Marina Abramović. Her artistic trajectory began with poetry, and, interestingly, in her first performance she included some of her verses. In *"Lo voy a gritar al viento"* (I'm going to shout it to the wind) (1999), she reads poetry out loud and tosses the pages in the air as she hangs from the arch of the Edificio de Correos, a historical building located in a very busy area of Guatemala City (Goldman 2006). For Galindo, this performance was very traditional, because "a pesar de haber contado con la presencia de muchos espectadores, su participación fue pasiva. Se paran en la calle y me observan mientras leo mis poemas en voz alta" (Galindo 2012; Despite the presence of many spectators, their participation was passive. They stood on the street and observed me while I read my poems out loud). Since then, her approach to performance has changed to include audience members, volunteers, or paid participants. This is an important aspect of her work, as part of the experience she evokes needs to be completed, which means that her audience not only understands the dialogue she initiates, but is able to supply the rest of the script: "Construyo las imágenes y espero a que el público termine de elaborar la imagen a través de sus acciones. Y si no participa el público, no termina de crearse la imagen" (Galindo 2012; I construct the images and wait for the audience to finish creating the image through their actions. And if the audience does not participate, the image is not completed).

Part of what allows these performances to be completed is their accessibility, as she presents them in spaces where her work is both expected and unexpected: on the streets, where violence and its traces are commonplace, but also where art seems out of place. In other instances, whether intentionally or not, the work is presented in sites that have a historical significance that ties in to her work – though ultimately it could be argued that violence, racism, and impunity are so pervasive that few places in Guatemala remain untouched by them – and make the performance that much more complex. What is also notable is that central to her work is a female body where violence is enacted, represented, and contested. With it, she sustains the gaze of her viewers and challenges them to reflect on the violence inflicted upon the female body by a society that is patriarchal and has been shaped by thirty-six years of war and more than five centuries of colonialism. Also of importance is the fact that the performance presented to the spectator, though undoubtedly denunciatory, relies only on the memory, and in some instances the archive, of the violence that is insinuated or at times re-enacted. In other words, while evoking the unspeakable, the performance does not repeat the initial experience of violence, thus providing, at least in theory, a safe space in which the spectator-witness can reflect on her own past experience as a victim, a witness, or both.

For Diego Sileo, a curator of Galindo's work in Milan in 2014, "[her] performances challenge customary cognitive strategies, generating a disturbing proximity between listening, witnessing and experience" (2014, 16). But no matter the distance between the real experience and the memory that Galindo activates, her work achieves a visceral response,[18] which, as discussed earlier, is key in performance. During a presentation of her latest work, *Culpable (Guilty)* (2015), a spectator reported that she began feeling a stirring in her chest, "a sentir algo dentro del pecho" (Prensa Comunitaria 2015). *Culpable* was presented in the colonial centre of Guatemala City, and it consists of a choral piece of only one word, *culpable*, sung for over an hour (Mejía 2015).[19] The melody, intricate and complete despite the single lyric, commemorates two years of Ríos Montt's guilty verdict on the charges of genocide, a ruling that has been challenged, overturned, and altogether denied. But more importantly, through this performance, Galindo links the past and the present, highlighting the validity of the charge of genocide despite the continued tergiversation of justice, and through the choral voices, delivers a performative utterance of a *public* (and unofficial) guilty verdict.

In *¿Quién puede borrar las huellas? (Who can erase the traces?)* (2003), another piece that deals with memory and forgetting, Galindo walks from la Corte de Constitucionalidad (CC; Constitutional Court) to the National Palace. Dressed in black, she carries a large basin filled with human blood, stopping occasionally to dip her feet in order to leave a trail of bloody footprints. The walk takes approximately forty-five minutes, and in the final stages she faces a group of police officers customarily lined up outside the palace (Goldman 2006). This walk was meant as a tribute to the victims of the war and her rejection of Ríos Montt's presidential candidacy in 2003: the distance that the performer covered was the same that Ríos Montt would have symbolically travelled if he were elected president. The bloody footprints in Galindo's *¿Quién puede borrar las huellas?* bring three temporalities together: the recent past, a time when Ríos Montt was the de facto ruler (1982–3) and a main leader of the genocide; the present (2003), a time when Ríos Montt vied to take the helm once again, despite the Supreme Court's initial ruling;[20] and a possible future that would be informed by the events (and actors) of the recent past. These temporalities intersect with the help of Galindo's body in a process in which each footprint is used to write a story that can be read by onlookers. Interestingly, for Galindo, in *¿Quién puede borrar las huellas?* memory comes before performance,[21] showing that what her work first activates in the spectator is not an aesthetic appreciation, but a recognition of a shared experience that is embodied in the bare feet, the blood, the silence, the anger, and in the seemingly vulnerable body of a woman. The transposed terms of her work, where memory comes before performance, are facilitated by avoiding the formal gallery space, but also by the immediacy of the recognition or self-identification by the spectator. In other words, the memory has not been eradicated, as it resides on the very surface and is easily accessible. The artist's handling of the memory of violence is reminiscent of the public's own: traces are left behind as life is resumed. After planting the two final footprints and leaving the large receptacle containing blood at the feet of the unsuspecting police officers guarding the National Palace, Galindo returned to the everyday: "I quickly walked across the street, washed my feet off in the park fountain, got something to eat, and then went back to my job that afternoon" (Goldman 2006).

For Galindo, the body is another site where memories reside and where they can be deployed. She often presents her spectator with a body full on, which she conceives of as communal, public, and necessary (Galindo 2014), in order to generate a reaction. In appearance, this body is petite, fragile, and female. This is especially meaningful, when she purposefully imbues it with the marks of violence (cuts, blood, apparent death, etc.) and appears before her spectators, challenging them to search for and face the unspeakable, the language that their own bodies understand but subsume in an effort to continue moving forward. As Joseph Roach explains, performance (more specifically, its genealogy) relies on "expressive movements as mnemonic reserves, including patterned movements made and remembered by bodies, residual movements retained implicitly in images or words (or in the silences between them), and imaginary movements dreamed in minds not prior to language but constitutive of it" (1996, 26). Galindo's work depends on the spectator's recognition and even self-identification with the body with which she presents them, a repository of memory, as she explains in an early artist's statement: "Mi cuerpo no como cuerpo individual sino como cuerpo colectivo, cuerpo global. Ser o reflejar a través de mi, la experiencia de otros; porque todos somos nosotros mismos y al mismo tiempo somos los otros. Un cuerpo que es, entonces, el cuerpo de muchos, que hace y se hace, que resiste y se resiste" (Olivares 2011; My body not as an individual body but a collective body, a global body. To be or to reflect through me, the experience of others; because we are all ourselves and at the same time we are others. A body that is, then, the body of many, that does and that is made, that resists and is resisted).

In her performances, audiences are confronted with a female body that carries with it the memory of a violence that take them back to the former, and in many cases to the current, Guatemalan state. In *Mientras, ellos siguen libres* (While they maintain their freedom) (2007), a naked and pregnant Galindo appears with her hands and feet bound with umbilical cords to a bed, lying in the same position in which soldiers held Indigenous women in order to rape them.[22] The documented performance is accompanied by a description that includes fragments of testimonies given to the CEH by two women who were raped during the war (Goldman 2006). *Mientras, ellos siguen libres* is a clear reminder of the violence perpetrated against women as well as a denunciation of ongoing impunity; the great majority of these cases have yet to arrive at a court of law, or even be officially recognized by the state as part of a broader strategy of genocide implemented by the state during the war. Though this particular performance was presented in the Edificio de Correos (Central Post Office), presently a cultural centre, its "placeness" is of great importance.[23] This structure, like many other important state buildings, was erected under president Jorge Ubico's dictatorial regime from 1931 to 1944, inspired by an ideology of progress and racial positivism. In fact, according to Marta Casaús, the practices of discrimination and genocide that are prevalent in Guatemala today can be traced back to the Ubico years.[24] The precepts of *blanqueamiento* (whitening) and *mejoramiento de la raza* (improvement of the race) are key in this scheme, and they have been used as justification for the objectification and violation of Indigenous female bodies, as Diane Nelson suggests, and contribute to the belief that the Indigenous female body is open and "rightfully" accessible to Ladino men (1999, 221).

When Galindo displays her body in such a manner, it is not difficult for audience members to fill in the blanks and reflect on the army's tactics to denigrate and exterminate a people, and on the inability of the state to bring to justice those who committed these crimes. More importantly, it shows the state's continued complicity in a system that condones racism and violence. The "placeness" of this performance further complicates the denunciation, as it takes the viewers to a symbolic origin of the country's official discourse on race, and with her body, that of a Ladina woman,[25] she shocks them into recognizing the humanity and familiarity of her body as it lies there, bare, vulnerable, and female.

Perhaps the most interesting but unsettling aspect of Galindo's work is that in performances where she relies on her body, she presents her viewer with a living body in the midst of being subjected to violence. To put it another way, rather than showcasing a beginning and an end, or a cause and an effect, Galindo shows violence *in media res*. This can be linked back to Galindo's strategy of requiring her viewers to fill in the blanks to complete the performance, but it also adds another layer to the question of memory. Not only does the artist offer a comment on the pervasiveness of violence, its effects, and continued presence in the country, but she also inserts her viewer into the narrative, coercing her to participate. The viewer can turn away, but at that moment her decision to do so has acquired the significance of a moral choice: to witness or not to witness; to empathize or not empathize; to remember or to forget. Galindo's persona, as both victim and witness, offers her spectator a role in the performance, transforming her into a witness, torturer, or simply a good samaritan. Giulia Casalini observes the witness-making effect in Galindo's work: "By performing violence and its traumatic memory on her body, Galindo is not only the witness of her own trauma but she also uncomfortably engages the spectator, who becomes witness at her/his turn. The public therefore becomes an active element in the performance, thus refusing the role of a passive spectator: activated by an ethics of responsibility the audience moves from the position of spectator to performer" (2013, 31).

In a more conceptual piece, *El peso de la sangre (The weight of blood)* (2004), performed in Guatemala City's main square, the Jardín Central, Galindo repeats the immediacy of violence as she presents us with a single signifier of it: a litre of human blood that slowly drips onto her head and body. Though violence itself is seemingly absent, the slow drip of fresh blood suggests the "hereness" and the "nowness" of it, and the presence of a body is a reminder of its human cost. And again, the artist situates her piece in a significant historical and political centre that has much to do with the spilling of blood in the country over the past half a century. As passersby look at the performance, though they many not be directly involved, they cannot escape its sociopolitical significance as a result of its "placeness" and the signifier of violence before them. Audience participation in *Mientras ellos siguen libres* and *El peso de la sangre* is much more passive than in other performances, but it benefits from a basic understanding of Guatemala's past, its present, or both. The latter performance, however, being abstract in its representation of violence, and relying less on the audience, invites the viewer to safely reflect upon piecing together a narrative through the visual cues (place, blood, female body) provided. Though most of Galindo's pieces elicit an "ethics of responsibility," as Casalini suggests, or utmost abjection, as Lavery and Bowskill argue, some of her work, less direct and less onerous for audience members, allows for a safe reflection on violence.

Rosa Chávez: Performing Resilience

On a busy Saturday, on 1 August 2009, during the Festivalote arts festival, Rosa Chávez walked ceremoniously along Santander Street in Panajachel and read her poetry out loud. Her actions were part of a performance used to present her second poetry book, *Piedra/Ab'aj* (Stone) (Chávez 2009). The street on which she performed is of particular importance because, as the main street in Panajachel, it serves as a type of portal for tourists and inhabitants alike, who must pass through to more readily access the outlying towns and villages of Lake Atitlán, a predominantly Tz'utujil and Kaqchikel area that draws a great deal of tourism. Wearing red pants and a red *huipil* (traditional woman's blouse), Chávez walked barefoot, following a path of yellow petals, towards a ceremonial space. The entire walk was done alongside her double, another Indigenous woman who was dressed like her and with whom she simultaneously read poetry in Quiché-Maya and Spanish. The two women moved through the main street by making their way amidst the crowd of tourists and Atitlecos, who merely observed, moved aside, or followed along. They both requested the permission of the "guardian spirit of the path" to pass through:

Dame permiso espíritu del camino
regálame permiso
para caminar
por este sendero de cemento
que abrieron en tu ombligo
por esta autopista de viento
que corta el silencio (Chávez 2009, 14)

Give me permission guardian spirit of the path
grant me permission
to walk
along this cement trail
they opened in your navel
along this road of wind
that cuts through silence

Once they reached the ceremonial space, the poet and her double each knelt on a wicker step stool facing each other, and continued reading, but this time alternating poems.

Rosa Chávez, a Quiché-Kaqchiquel-Maya woman, writes a poetry that is performative and permits the writer and her reader to engage with notions of gender and ethnicity, as she sets out to defy her reader with a body that is sensual, sexual, female, and Indigenous. Like Galindo, the Maya poet uses the female body explicitly and unapologetically to present an artistic project that extends well beyond the written word in order to question the validity of a deeply racialized patriarchal system. Her incursion into poetry is fairly recent, and she is one of very few Indigenous women writing in Guatemala today.[26] That day in August, when Chávez took to the streets to present her poetry collection *Piedra/Ab'aj*, she relied on performance to feature a femininity that does not conform to traditional representations and instead is complex, defiant, and urban. By means of her double, the poet also played with notions of Mayanness and Maya cosmogony, setting the stage for a poetic body that shifts power and rearticulates a contemporary indigeneity from the perspective of a Maya woman. The effect of her actions was heightened as she transformed her public into witnesses when she took on the rearticulation of Indigenous identity and reconfigured the space.

Chávez's choice of dress is key in her performance of Mayanness. The act of pairing a huipil, one of the most iconic articles of Indigenous clothing, with trousers, defies convention for both Ladino and Indigenous viewers. The *traje típico* (traditional dress), which includes both *corte* (skirt) and huipil for women, stands as an emblem of Guatemalanness for Ladinos and as a symbol of tradition as well as a locus of cultural resistance and determination for Indigenous people. When she alters the traje, Chávez contests both conceptions, presenting a new form of wearing it that is not tied to the dominant culture or to tradition but is her own.[27] From a gender perspective, the alteration of the traje is particularly defiant, as it is mostly women who weave and wear it, and they are by and large responsible for the protection and production of Maya culture and tradition. The symbolism of the transmutation of her dress cannot be denied: donning a pair of trousers shifts the expected performance but also brings up questions of identity. As Diane Nelson showed in her analysis of the jokes about Rigoberta Menchú, in the choice of dress of an Indigenous woman – whether or not a corte is worn – lies an anxiety of authenticity for the ladino viewer (1999, 175).

Alongside Chávez's *traje* (a)*típico*, is the question of her double. The poet's double can be read as a contestation of tradition that sees her lay bare her *tijax*, which represents obsidian stone. Menchú, in her testimony, describes the nahual as a presence or energy that from the time of birth accompanies each Maya person: "Every child is born with a *nahual*. The *nahual* is like a shadow, his protective spirit who will go through life with him" (Burgos-Debray 1984, 18). Menchú refuses to reveal her nahual, claiming that it must be kept secret in order to preserve her culture, a move that is understood by Arturo Arias as discursive strategy and by Doris Sommer as incommensurable cultural difference (Arias 2001, 79–80; Sommer 1991, 34). In *Piedra/Ab'aj* Chávez addresses this duality, as the poetic voice describes her birth as the birth of two hearts, "[d]e dos corazones salí al mundo" (from two hearts I came into the world) and in a transformative state, a penchant for revealing secrets "me vuelvo hierba que devela secretos" (56; I become grass that unveils secrets). Similarly, in public, on the streets and in interviews, the poet does not feel the need to protect *her* secret as she gives it a body and puts it on public display, no longer fearing the dominant culture, because her poetry is "made of obsidian" (Hernández 2010). For example, in a feature in the cultural supplement of the *Siglo 21* newspaper, *Magacín*, Chávez appeared facing the camera head on, holding a bloody heart. It was in this feature that she first spoke openly about her nahual.

Piedra/Ab'aj also introduces Kame, a nahual of death and rebirth, to which, in an unexpected turn, the poetic voice assigns a gender and a human name: "Elena Kame/crisantemo/flor de muerto/florecita de azufre" (24; Elena Kame/chrysanthemum/flower of the dead/little flower of sulfur). Chávez's subsequent poetry collection *Quitapenas* (2010) is dedicated to this nahual, a force that is key in the poetic voice's trajectory: "Para el Kame de los círculos infinitos, para el / Kame de sonajas en las manos y los pies, para el / Kame que me protege y que me ha enseñado a / danzar con la muerte la danza de la vida" (2010, 7; To the Kame of infinite circles, to the / Kame with rattles on hands and feet, to the / Kame that protects and that has taught me to / dance with death the dance of life). Interestingly, this is the only poem that is translated into Quiché in the entire collection. *Quitapenas* is a series of poems dealing with pain and suffering that activates those same feelings in the reader. But rather than simply mimicking them, it initiates a process of healing and renewal. The title of the book itself is significant, as it refers to the traditional small handcrafted dolls made of cloth, paper, and wire to which one can whisper one's worries so that they can be taken away during the night, thus bringing peaceful sleep. The effect is performative: it consists of facing or naming the worry in order to begin to deal with it. Chávez's poetry fulfils a similar role and it also incorporates healing into the performative. In the words of the author, *Quitapenas* "se trata de un poemario que reflexiona sobre la restauración de un estado emocional, de sanar, de quitar las penas … pero desde la perspectiva maya, donde acabar es iniciar un nuevo ciclo" (Hernández 2010; it's a poetry collection that reflects on the renewal of an emotional state, of healing, of doing away with worry … but from a Maya perspective, in which to end is to begin a new cycle).

In her blog, Chávez returns to Kame, who appears in human form, and has an urban experience in which she is subjected to a violence and a forgetting that pervades Guatemala City:

> Elena Kame tiene la enfermedad del susto,
> la asaltaron en la 6ta. avenida y su espíritu quedo allí,
> tirado, machucado,
> en medio del tráfico,
> esperando que alguien lo recoja. (2008)

> Elena Kame is sick with fright,
> she was robbed on 6th Ave. and her spirit remained there,
> tossed, trampled,
> in the middle of traffic,
> waiting for someone to pick it up.

Chávez's exploration of the urban landscape has a precedent in her first poetry collection, *Casa Solitaria* (Solitary home) (2005), where the poetic voice opens with the experience of an Indigenous woman who has recently migrated to the city and is immediately confronted by the disintegration of her family, hunger, domestic labour, and racism:

> hace un mes
> vine a la capital
> mi tata nos abandonó
> y en la casa el hambre dolía,
> yo trabajo en una casa
> (la señora dice que de doméstica
> aunque no entiendo muy bien que es eso,
> me dieron un disfraz de tela). (5)

> a month ago
> I arrived in the capital
> my dad abandoned us
> and at home hunger hurt,

I work in a house
(the lady says that I am a domestic worker
but I don't really understand what that is,
they gave me a costume made out of fabric).

Later on, the poetic voice in *Casa Solitaria* conceives of herself as a type of machine that wanders the streets in the capital, visiting abject spaces, and absorbing the urban experiences of a cast of rejects (prostitutes, drug addicts, maids, *indios*, thieves) that inhabit therein:

Desenchufar
cortar la energía del profundo generador
que me mueve objeto máquina
por la calle
me veo en el ladrón
me intimida, sabe que sé su secreto
perturba el aliento de la ciudad (36)

To unplug
to cut the power of the deep generator
that moves me, machine object,
through the street
I see myself in the robber
he intimidates me, he knows I know his secret
the breath of the city perturbs

This process of moving through the city requires that the poetic body not only travel along the streets and through those spaces, but in doing so, also activate for the reader the experiences of those "others." The poem begins in the first person but slowly assumes a collective identity that speaks of violence and crime as a common social ill. Here, when Chávez switches to the first-person plural, she recalls an everyday sentiment that Guatemala City dwellers are all too familiar with: anger and frustration, and fear and anxiety, compounded by a feeling of powerlessness that resembles that of "animales atados a un poste" (36; animals tied to a post) in a city with one of the highest violent crime rates in Central America. This collective identity is made possible through a process of recognition and self-identification much like what occurs in Galindo's work.

Returning to Santander Street and using De Certeau's concept of "walking in the city," in which the pedestrian has the power to create and transform the urban space through her own movements, I would like to suggest that the poet and her double approach the street in a manner that is both assertive and highly transformational: they take possession of a street that is defined by its everyday commerce and transit of tourists and locals and, for a short while, they convert it into an urban ceremonial space. In fact, a number of transformations take place: the dynamics of the everyday of Santander Street are suspended, onlookers are turned into members of an urban congregation of sorts, the street becomes a sacred path, and the poet leads in a renewed ceremonial practice. The performativity of the verses read along the way is also key in the reconversion of the street because as she walks and transforms, she utters the words that name and complete the process.

In the poem that accompanies the trajectory, "Dame permiso espíritu del camino" (Give me permission, Spirit of the Road), images of the urban – a cement trail and a highway – are presented as painful intrusions upon the path, carved into its very core, "que abrieron en tu ombligo" (2009, 14; that they opened in your navel).[28] Here, the image of the navel is of special importance, given that in Maya cosmogony the navel is a place of origin as well as a centre where cardinal points intersect, and from which the different realms can be accessed. The violence suggested in the action of opening a cement road through a navel is reminiscent of the violence inherent in the various modernization projects in the country, characterized for their impetus to tame and control nature, particularly in the zones inhabited by Indigenous people, and deemed necessary for economic and social development. But at the same time that she acknowledges the violence, the poetic voice shifts the dynamic by attributing great agency to the inhabitants of the natural world (birds, plants, animals, and stones):

Dame permiso espíritu del camino
[...]
permiso también a ustedes
pájaros que rompen el tímpano del acero
permiso piedras
permiso plantas
permiso animales que resisten en la neblina. (2009, 14)

Grant me permission, Spirit of the Path
[...]
permission from you too

birds who break the steel drum
permission, stones
permission, plants
permission, animals who resist in the fog.

This poem employs five apostrophes, each addressing a distinct organism (living and non-living) and signalling that each one deserves respect, while showing that there is a dialogue between them and the speaker. This relates to the intrinsic notion of balance in the Maya world view that benefits from communion with nature: "For many Maya today, continued human existence is predicated on the maintenance of a cosmic balance that both affects and reflects earthly conditions" (Fischer 2001, 147). The seemingly innocuous act of asking organic entities, which are seldom associated with modernity, for permission reveals a divestment and redistribution of power, as well as rejection of anthroprocentric world views. Birds are said to possess the capacity to break "el tímpano del acero," an allusion to fabricated sounds and materials associated with the idea of industrialization and, in the case of the "tímpano" or *timbal* (a drum often used by musical bands and military ceremonies) militarization, or in its literal sense, eardrum. Meanwhile, animals stand by in the fog resisting forgetting and extermination, and volcanoes sing and hills speak:

dejáme por favor regresar a mi casa,
antes de que los volcanes canten,
antes de que el discurso de los cerros
escupa en nuestras bocas. (2009, 14)

let me return to my home,
before the volcanoes sing,
before the speech of the hills
spits in our mouths.

The poetic voice sets in motion a decolonizing act that rejects the status quo and transmutes power and authority, showing that humankind is part of a much broader system that incorporates the natural world. From this perspective, birds, paths, plants, and stones are given the prominence that they are due, and hills and volcanoes, endowed with their own speech, enter the political arena. This is reminiscent of Marisol de la Cadena's call for a recognition of sentient "earth beings" as political actors in her study of Indigenous activism in the Andes (2010, 341–2). For de la Cadena, this new Indigenous politics deftly challenges the manner in which the nation state functions and overturns an entire system that threatens the livelihood of Indigenous people. Similarly, Chávez's verses give prominence to those "earth beings," transforming the space that she traverses and calling to alter the relations of power and coloniality that exist therein. At the same time, she shows that those entities have withstood the test of time and the violence inflicted upon them, and she, like they, stands resolutely. Nevertheless, it would be naive to assume that such experience comes without a cost: the poetic voice warns us that anger exists within, "rabia que desorbita mis ojos" (anger that alters my senses), and that she must reach home before all will is lost, "dejáme pasar/que mi voluntad no se pierda" (let me pass / so that I don't lose my will).

In another poem, a similar call for breaking the colonial binds is issued by a new poetic presence, a grandmother, who transmits the memory of forced labour and violence, but who also appropriates and passes on the product of her suffering: "Esta carretera también es tuya mi'ja / la pagó la esclavitud/el tiempo bajo el padre sol" (*Piedra/Ab'aj* 68; This road is also yours dear / slavery paid for it / time spent under father sun). This appropriation signals a recognition of the rights of the individual and the community, at the same time that it condemns those who put into motion the idea of progress and ordered the construction of a road. In this poem, as in Chávez's performance, the road, a seemingly neutral zone, becomes a space where power is contested, memory activated, and (neo-)colonial relations upturned. Indigenous agency is key, as is the notion that Indigenous people are resilient.

The notion of resilience appears throughout Chávez's poetic body. For example, the implications of how she chose to perform her ethnic and gender identity in the *Magacín* feature, discussed above, are numerous, but at least one message is made clear in the subheading of the piece: "latir sin descanso" (Hernández 2010; to beat without pause). In the feature, the various images of Chávez can be divided into two groups: the poet dressed in white, holding a bleeding heart, and the poet wearing the traje. The image of the heart and the line "latir sin descanso" are direct references to the second poem in *Quitapenas*:

Nos quitan la cabeza y el corazón sigue latiendo
nos arrancan el pellejo y el corazón sigue latiendo
nos parten a la mitad y el corazón sigue latiendo
beben nuestra sangre y el corazón sigue latiendo
estamos criados para latir sin descanso. (Chávez 2010, 12)

They behead us and the heart keeps on beating
they skin us and the heart keeps on beating
they split us in half and the heart keeps on beating
they drink our blood and the heart keeps on beating
we have been raised to beat without pause.

The violence that each of these acts suggests is challenged by a refusal to be exterminated as a people, a tension that is echoed in the use of a first-person plural, a victimized *we*, and a third-person plural, a victimizing *they*. Each violent act named in these verses (decapitation, skinning, slicing, drinking blood) is not attributed to any one group, but the reader can deduce, from the sadism that these actions require, that the perpetrators are the *kaibiles*, the Guatemalan counterinsurgency special forces that committed some of the most heinous acts during the war. The violence occurs in the present tense and is reminiscent of Galindo's representation of a *violence in media res,* which points to an open wound, an unresolved situation that continues affecting an entire people. While the mention of these violent acts reaches into memory, be it personal or collective, the notion that it brings to the forefront is that of resilience: despite the violence and the cruelty, the heart keeps on beating, steadily until the very end, at which point it does not stop but continues. The way in which Chávez deals with violence in her poetry is similar to Galindo's handling of the theme in her performance, but unlike Galindo, she evokes a different unsayable element, one that offers a sort of hope: not fear or pain, but the strength to endure and overcome even the most extreme circumstances.

Galindo and Chávez: Poetry and Performance

Implicit in Galindo's and Chávez's real and figurative journeys through city streets are questions of what it means to deal with collective memory in twenty-first-century Guatemala. Their ceremonial walks to the National Palace, through Panajachel, or on a sacred path transform space by challenging the roles they have been assigned and the histories they have been told, and by disrupting silence and invisibility. In their path they leave behind audiences that have been stirred and prompted to look anew at the female body, the "profundo generador" (Chávez 2005, 36; complex generator) that propels them forward and that functions as a site of reflection. With each footstep (and for the duration of their performance) they take the street and make it a text to be read against unresolved socio-cultural and political realities, and personal and collective experiences and memories. In *¿Quién puede borrar las huellas?* Galindo lends a body to the memory of violence and death by tracing it in human blood, thus reviving it. For her part, Chávez, through her various walks, both physical and metaphoric, transforms the street by renewing it as a Mayan space that is not anthropocentric but part of larger narrative that, despite being met with violence, has endured.

Galindo and Chávez have worked together on two projects, with a layer of complexity that is made possible by their particular approaches to violence and memory: *Hermana* (Sister) (2010) and *Inmonumental* (Unmonumental) (Galindo and Chávez 2013). Though Galindo has worked on issues of ethnicity, she recognizes that they are limited in their scope, given that she is Ladina and she cannot see them through any other lens. For example, to put together *La conquista* (The conquest) (2009), Galindo explains that she placed an ad in the paper, offering to buy the hair of Indigenous women so that with it she could create a hairpiece sculpture (Galindo 2012). In the initial stages Galindo became aware of how truly complex and lopsided Ladino-Indigenous relations really are in Guatemala: first she thought that poor women themselves would be selling their hair out of necessity, but soon after placing the ad, it became evident to her that others saw an opportunity. As a result, Galindo altered her strategy, as she feared for the safety of Indigenous women:

> Eso fue mucho más allá del "no encuentro" entre dos grupos; se trataba de un abuso completo. Me di cuenta que podría estar provocando un abuso mayor al querer comprar cabello de mujeres indígenas. O peor: cualquier desgraciado podría asaltar a una mujer en la calle y cortarle el cabello con tal de venderlo. Decidí comprarles el cabello únicamente a las mismas mujeres, eso me hizo reflexionar sobre muchas cosas. (Galindo 2012)

> This went much further than a "non-encounter" between two groups: it was complete abuse. I realized I could be contributing to a greater abuse by wanting to buy the hair of Indigenous women. Or even worse: any bastard could mug a woman on the street and cut her hair with the intention of selling it. I decided to buy hair only directly from the women themselves. This made me reflect on many things.

Chávez, for her part, though she also deals with violence, concentrates in her work on the memory of a people and their resistance: "Un aquí estoy, como desafío al mundo que se desmorona, ayer por ser el día en que todo nace y muere" (Hernández 2010; An "I am here," as defiance to a world that is falling apart, yesterday as the day that everything is born and dies). At the same time, the Maya poet resists traditional notions of Indigeneity and proposes new forms, as discussed above.

Hermana is the first piece in which both women come together, and it is especially interesting because the two artists share an intimacy that is explored in the title of the performance itself: they are well known *comadres*. *Hermana* is shown in the format of a triptych video, each section lasting thirty-two seconds; in it Chávez appears dressed in traditional Maya garb, as she sets out to slap, spit on, and hit Galindo repeatedly with a stick: "Mi cuerpo ladino es abofeteado, escupido y castigado por una mujer indígena guatemalteca" (My Ladino body is slapped, spat on, and punished by a Guatemalan Indigenous woman).[29] Galindo, for her part, remains passive, wearing jeans and a black long-sleeved T-shirt, this last article of clothing removed in the third part of the performance so that the caning is received directly on her bare back. Here we see a shift of hierarchies: the Ladino body is now the recipient of centuries-old violence at the hands of a subaltern subject. This performance is difficult to take in, not only for the violence presented to the viewer, but for the close relationship between the two women suggested by the title: despite being "sisters," there is a power differential that, in this case, is turned on its head. The reversal that takes place adds an impossible dimension to the violence in the country: an Indigenous woman wielding power, dominating a Ladina. In this performance Galindo and Chávez disrupt the naturalized order, and they prompt the viewer to reflect on the normalization of violence inflicted upon Indigenous bodies. The incongruence that confronts the viewer, rather than merely contesting expression of fraternity or female solidarity, is a questioning of the neoliberal multicultural discourse and the status quo itself.

Inmonumental is a much broader project, consisting of a set of performances by different artists in a new artistic space Concepción 41, an old convent in Antigua, Guatemala. To inaugurate the space, artists set out to resignify the space by playing with the memory imbued in the convent (Concepción 41 2012).[30] Galindo and Chávez carried out their parts separately but present them together in a video as *Immonumental*. In Galindo's piece, *caminos* (paths), she lies wrapped in a white shroud in the middle of a desolate field. Her body, or rather a large bundle, is tied with four white cords that have been unfurled by four different women and distributed in four directions. Upon stumbling upon one of the cords in the street, the viewer may follow it and eventually come across the body, at the place where it has been "dumped": "Mi cuerpo permanece escondido dentro de un matorral. El cuerpo es un bulto, amarrado con cuatro hilos que son sacados por cuatro mujeres hacia el espacio exterior. Los hilos hacen dibujos por las calles de Antigua Guatemala. El público debe de seguir los hilos para encontrar el cuerpo" (Galindo n.d.; My body remains hidden in the brush. The body is a bundle tied up with four strings that are drawn out by four women into the open. The strings trace drawings through the streets of Antigua Guatemala. The audience must follow the strings to find the body). Simultaneously, Rosa Chávez is shown on the split screen approaching an abandoned pool that is filled with debris. She does so as she swings an incence burner and makes her way atop the heap of debris, to then begin reading a poem about *los cuatro caminos*, the four intersecting paths of Maya mythology that link the cosmos (Tedlock 1996, 340). She talks about pain, death, and memory, and she specifically asks the four corners to help with the process of healing: "Cortaremos el hilo de la historia de dolor en nuestra sangre, lavaremos nuestra ropa en un río, la golpearemos fuerte, la somataremos en las piedras para curarnos del susto. Que el agua se lleve lo que nos amarra, que las piedras guarden la rabia que atraviesa nuestros ojos" (We will cut the string of the history of pain that runs in our blood, we will wash our clothes in the river, we will strike them hard, we will beat them against the stones to cure ourselves from fright. May the water take away that which ties us down, may the stones keep the anger that eclipses our eyes).

The four intersecting paths that Chávez invokes are transubstantiated in the cords that lead to Galindo's body, a body that represents death but is not dead. Like the hero twins of the *Popol Vuh*, in *Inmonumental* Galindo and Chávez summon death, not to fear it, but to face it head on. Chávez calls for purification, for a healing process that deals with the anger and the pain and that allows for rebirth. In this piece, pain is a memory retained the body, while *susto* (fear) is a trauma that needs to be cured; a body, like a memory, needs to be found and retrieved to arrive at a place where one can begin.

Conclusion

The case of Rodrigo Rosenberg brings to light the conveniently imaginary rule of law under which Guatemala has operated for at least half a century. It helped show that to elicit a response from a government that legitimizes its despotic rule, denies genocide, and relies heavily on political theatrics, a performance that resists oblivion is required. I departed from the understanding that the state relies on its political performativity for the authentication and configuration of its power, a fact that is well known by the majority of Guatemalans, given that the state takes this one step further and into the realm of the histrionic. Post-conflict governments thwart the justice and reconciliation processes by deploying a repertoire that includes the continued theatrics of politics and the institution of violence, both under the protective veil of near-absolute impunity. A result of the state's heavy-handed political theatre is the obfuscation of the real, albeit sometimes unsayable, effects of violence, during and after the conflict. I have argued that Galindo's and Chávez's work confronts the Guatemalan public and the state by refusing to forget, by showing what cannot be said, and thus activating what has inevitably become part of the archive in recent years. The processes of healing and reconciliation can be jump-started by retrieving memories of the unspeakable through a safe medium, such as art, but art alone is not enough, because the erasure of historical memory or its archival "shelving" can occur officially, as sanctioned and upheld by agents of the state. The work of these two women provides Guatemalans with a place to deal openly with the experiences of violence, as it sparks a conversation outside officialdom, where it continues to be silenced. While they may not offer a solution, they do offer a place of reflection where healing can begin.

Galindo's well-known body of work touches on state violence, collective memory, and trauma, and Chávez's possesses an important dimension that speaks directly to the ethnic violence that characterized the conflict and has shaped ethnic relations ever since. Over the past decade, their work has made explicit what the state attempts to conceal: a violated body that ails and that demands healing, a body that is political, female, Ladino, and Indigenous.

NOTES

1 Translations from the Spanish are by the author.
2 For a detailed account of the case, see Grann (2011).
3 In Guatemala *Ladino* is a term used to describe those who do not identify as Indigenous and is based primarily on cultural difference (language, dress, tradition, custom), though race also plays a part. It most closely resembles the concept of mestizo that has wider currency in the rest of Latin America.

4 An agreement between the UN and the Guatemalan government identified some of the groups that currently undermine the country's legal system and perpetuate corruption and impunity as Cuerpos Ilegales y Aparatos Clandestinos de Seguridad (CIACS; illegal groups and clandestine security organizations). These CIACS had their beginnings during the war, as the state's counter-insurgency force that, according to the centrist think tank ASIES, in post-conflict Guatemala has evolved: "[Los CIACS] han ido transformándose, mimetizándose y adaptándose a las circunstancias actuales, convirtiéndose en maquinarias muy sofisticadas al servicio de estructuras de crimen organizado, que debilitan el Estado a alimentar la corrupción a todo nivel y aprovechándose de las instituciones políticas y de sus fondos para generar inmensas ganancias con toda impunidad" (ASIES 2015; CIACS have transformed, mimicked, and adapted to current circumstances, becoming sophisticated machines at the disposal of organized crime that weaken the state by contributing to corruption at every level, and use political organizations and their funds to generate immense profit with absolute impunity).

5 According to official figures, in 2009, 99.75 per cent of violent crime went unpunished (Human Rights Watch 2011, 243).

6 The theatrics employed to sustain the state in Guatemala are well known, and citizens quickly recognize its features. In late June 2015, as CICIG brought corruption charges against different levels of Otto Pérez Molina's government, opinion pieces in *El Periódico* reflected on the high degree of fiction, spectacle, and theatricality under which the Guatemalan government operates, calling it a vaudeville (a type of musical farce) and remarking on a politics of "special effects" that relegates dialogue and ideas to oblivion (Chea Urruela 2015). The notion of a circus with the president as its conductor has also been evoked (Vela Castañeda 2015).

7 Though there have been many high-profile cases in the last two decades worthy of commentary, particularly given the manner in which they were handled, Rodrigo Rosenberg presents a unique case. Unlike many of the other cases, Rosenberg's murder was not politically motivated, and he, not the state, crafted the fiction. I chose this particular case because it helps demonstrate that to be part of the body politic in Guatemala, individuals must understand the language of the state, which requires a high degree of performance and spectacle.

8 During a televised address on CICIG's findings on the case, Carlos Castresana, head of the commission at the time, went so far as to declare that Rosenberg "was an honourable person.... He wanted to open up a Pandora's box that would change the country" (Grann 2011, 25–30).

9 The CEH visited over 2,000 communities and registered over 8,000 testimonies on the human rights violations committed during the war.

10 Guatemala is not unique in this respect. Other countries that have faced or are facing similar postwar processes are presented with the same quandary. In the late nineties, for example, as the Truth and Reconciliation Commission was just beginning its work in South Africa, *Memórias íntimas marcas* (Memories Intimacies Traces), a project that explored the deep effects of the Angolan War, set out to address the need for moving beyond the legal and the political, in order to delve into collective and individual experiences for a better understanding of the violence and gross human rights violations that took place. The three artists involved in the project, Fernando Alvim (Angola), Carlos Garaicoa (Cuba), and Gavin Younge (South Africa), saw their work as distinct but complementary to the TRC's own: "As citizens of the 'new' South Africa, we cannot afford to invest in placebo cures to the past. We need to explore our consciences and our complicity with recent history, deconstructing the legacies of apartheid. This cannot only happen 'officially' as it is currently through the Truth and Reconciliation Commission; it is an invested process which involves the individual and needs to be enacted on many levels as part of the process of establishing a way forward and recognizing that the future is complex, en-grained and marked with the traces of the past, the resonance of process" (artists' statement cited in Curiel 2006, 37).

11 Three weeks after CEH published its report, *Guatemala: Memoria del silencio/ Guatemala: Memory of Silence*, President Arzú's signature appeared in a one-page ad in a leading national newspaper, rejecting the CEH report's findings.

12 I would argue that this archive does not even constitute an official archive, as the state has denied the genocide and diminished the impact of the war upon Indigenous people.

13 Catherine Nolin in chapter 2 and Stephen Henighan in chapter 8 develop distinct, yet related, definitions of this term.

14 Statement cited in Amnesty International (2002, 5).

15 Unless otherwise indicated, Galindo's performances are archived in her website, Regina José Galindo.

16 According to the CEH, "La estimación de desplazados oscila entre 500 mil y un millón y medio de personas en el período de mayor afectación (1981–1983), sumando las que se desplazaron internamente y también aquellas que se vieron obligadas a buscar refugio fuera del país" (1999, 3:120; Estimates of displaced people range between five hundred thousand and a million and a half during the most severe period [1981–3], and they

include those who were internally displaced and those who were forced to seek refuge outside of the country). The Organización Internacional para las Migraciones (OIM; International Organization of Migration) reported that over 1.6 million Guatemalans were living abroad in 2010 (2013, 48).

17 During the war, the main actors that played with the ideas of truth and justice for their own benefit were the branches of the state that were in charge of enforcement and security. In *The Guatemalan Military Project*, Jennifer Schirmer discusses the importance of legalism in legitimizing state-sponsored violence and impunity, as power was transformed into law (1998, 125–8). Post-conflict, the result is a system that operates under the thumb of what a WOLA report has identified as *"los poderes ocultos"* (hidden powers): "La expresión *poderes ocultos* hace referencia a una red informal y amorfa de individuos poderosos de Guatemala que se sirven de sus posiciones y contactos en los sectores público y privado para enriquecerse a través de actividades ilegales y protegerse ante la persecución de los delitos que cometen. Esto representa una situación no ortodoxa en la que las autoridades legales del estado tienen todavía formalmente el poder pero, de hecho, son los miembros de la red informal quienes controlan el poder real en el país. Aunque su poder esté oculto, la influencia de la red es suficiente como para maniatar a los que amenazan sus intereses, incluidos los agentes del estado" (Peacock and Beltrán 2003, 5; The expression *hidden powers* refers to an informal and amorphous network of powerful individuals from Guatemala who use their positions and contacts in public and private sectors for financial gain through illegal activity and to protect themselves from being prosecuted for the crimes they commit. This represents an unorthodox situation in which the state's legal authorities formally hold power but, in fact, the members of this informal network control real power in the country. Even if their power is hidden, the network's influence is enough to hamstring those who threaten their interests, including agents of the state).

18 The visceral dimension of Galindo's work is hard to ignore, and some critics have focused on her use of an abject female body to better understand the artist's approach to denouncing violence. For more, see Lavery and Bowskill (2012) and Rodrigues (2012).

19 The choral piece was written and conducted by Vinicio Salazar and performed by Estudio Coral de Guatemala.

20 In May 2003, after Ríos Montt's nomination by the right-wing FGR party, the Tribunal Supremo Electoral (TSE; Supreme Electoral Court) rejected Ríos Montt's bid for presidential candidacy (an article in the Guatemalan constitution prohibits anyone involved in a coup d'etat from running for the presidency). He twice appealed to the TSE and later sought an

injunction with the Corte de Suprema Justicia (CSJ; Supreme Court), which was also denied. He then appealed to the CC and the request was also denied. After violent protests organized by Ríos Montt himself, the CC ordered the TSE to allow his candidacy to proceed (Schirmer 2003).

21 "My long walk of the bloody footprints was not initially understood as a performance, but every step was indeed understood as memory and death" (Goldman 2006).

22 According to the CEH, during the war 30,000 women were raped, 89 per cent of whom were Indigenous (1999, 3:24).

23 Since then, this piece has been shown in other venues but as a documented performance.

24 Specifically, these notions of racial positivism were promoted by presidential secretary and respected Guatemalan writer Carlos Samayoa Chinchilla as well as his peers. Years after serving Ubico, Samayoa Chinchilla became minister of the Instituto de Antropología e Historia (Anthropology and History Institute). For more on the subject of race in Guatemala, see Casaús Arzú (2007). Henighan and Johnson, in chapter 10, also discuss the historical roots of Guatemalan Ladino racism.

25 Galindo explains that her intention is not to represent or speak on behalf of Indigenous people but to portray the violence that has affected people in Guatemala: "En mi caso particular, en muchas obras *hago con mi cuerpo lo que les sucede a muchas y a muchos.*... Yo creo que sería poco ético hablar por una comunidad indígena, pues no me corresponde. Yo soy ladina (persona no indígena) y estoy muy consciente de ello. Así fui criada, mis padres fueron criados como ladinos. Por más que tenga la misma fisionomía y la misma sangre que ellos, no tengo la misma herencia cultural. No se es indígena maya porque se tenga la sangre maya. Si sos maya es porque te criaron con la cultura quiché o kaqchiquel, porque defendieron con la vida esa herencia maya. Yo no voy a venir a ponerme la etiqueta de 'indígena' porque ahora está de moda lo maya" (Galindo 2012, my emphasis; In my case, in many of my works *I do with my body what happens to many women and men*.... I think it would be unethical to speak on behalf of an Indigenous community, since it's not up to me. I am Ladina (a non-Indigenous person) and I am very aware of it. I was raised in this manner; my parents were raised as Ladinos. Although I share the same looks, the same blood as them [Indigenous people], I don't have the same cultural heritage. One cannot be Indigenous Maya because one has Maya blood. If you're Maya it's because you were raised in the Quiché or Kaqchiquel culture, because they defended that cultural heritage with their lives. I will not adopt an 'Indigenous' label just because it's fashionable now to be Maya).

26 Calixta Gabriel Xiquín, Maya Cu, Adela Delgado Pop, María Elena Nij Nij, Blanca Estela Colop Alvarado, Negma Coy, and Ixmucané Us can be counted in this very small group of Indigenous women writers.

27 Manuela Camus's *Ser indígena en Ciudad de Guatemala* (Being Indigenous in Guatemala City) is a well thought-out study that documents Indigenous identity in the capital. Through a series of interviews, she discovered that when Indigenous men ceased wearing the traditional dress it was more widely accepted than when women did so (2002, 314n8).

28 For the titles of the poems, I follow *Piedra/Ab'aj*'s format. Each poem is listed in the index using its first line as a title.

29 Description that accompanies the documented performance on the artist's website (Regina José Galindo).

30 From Concepción 41's website: "INMONUMENTAL gathers a group of artists that react upon [*sic*] a charged space and its monumental presence to explore other means of interpretation. It is a mobilization of the idea that monuments contain, represent, and fix, in order to advance through other senses of memory, politics and democracy. In an epoch of broken icons and fragile ideologies, this project promotes an experience in transit, where the intention is to imagine, while at the same time speculate beyond the walls of a monument."

REFERENCES

Amnesty International. 2002. *Guatemala's Lethal Legacy: Past Impunity and Renewed Human Rights Violations*. London: International Secretariat.

– 2012. "Guatemala: Impunity, Insecurity and Discrimination. Submission to the UN Universal Periodic Review, 14th Session of the UPR Working Group, October–November 2012." April. https://www.amnesty.org/en/documents/amr34/004/2012/en/.

Antze, Paul, and Michael Lambek, eds. 1996. *Tense Past: Cultural Essays in Trauma and Memory*. New York: Routledge.

Arendt, Hannah. 1991. "Organized Guilt and Universal Responsibility." In *Collective Responsibility: Five Decades of Debate in Theoretical and Applied Ethics*, edited by Larry May and Stacey Hoffman, 273–83. Savage, MD: Rowman & Littlefield.

Arias, Arturo. 2001. "Authoring Ethnicized Subjects: Rigoberta Menchú and the Performative of the Subaltern Self." *PMLA* 116 (1): 75–88.

ASIES. 2015. "ASIES pide permanencia de CICIG y reforma al sector justicia." Asociación de Investigación y Estudios Sociales (ASIES). 20 January. http://www.asies.org.gt/comunicado-de-prensa-2/.

Burgos-Debray, Elisabeth, ed. 1984. *I, Rigoberta Menchú: An Indian Woman in Guatemala*, translated by Ann Wright. London: Verso.

Calderón Bentin, Sebastián. 2014. "The Rosenberg Video: *Testimonio*, Theatricality and Baroque Politics in Contemporary Guatemala." *Identities (Yverdon)* 21 (4): 364–79. https://doi.org/10.1080/1070289X.2013.847372.

Camus, Manuela. 2002. *Ser indígena en Ciudad de Guatemala*. Guatemala: Facultad Latinoamericana de Ciencias Sociales.

Casalini, Giulia. 2013. "Feminist Embodiments of Silence: Performing the Intolerable Speech in the Work of Regina José Galindo." *Ex æquo* 27:27–41. http://www.scielo.mec.pt/scielo.php?script=sci_arttext&pid=S0874-55602013000100003&lng=en&nrm=iso&tlng=en.

Casaús Arzú, Marta. 2007. *Guatemala: Linaje y Racismo*. Guatemala: Guatemala F&G Editores.

Chávez, Rosa. 2005. *Casa Solitaria*. Guatemala: Ediciones Ermita.

– 2008. "Poemas desenvueltos, desatados." Santa Tirana (blog), 28 April. http://santatirana.blogspot.ca.

– 2009. *Piedra/Ab'aj*. Guatemala: Cultura Editorial.

– 2010. *Quitapenas*. Guatemala: Catafixia.

Chea Urruela, José Luis. 2015. "La política como entretenimiento." *elPeriódico de Guatemala*, 27 June. www.elperiodico.com.gt. Accessed 30 June 2015.

Comisión para el Esclarecimiento Histórico. 1999. *Guatemala: Memoria del silencio*. 12 vols. Guatemala: CEH.

Concepción 41. 2012. "Concepción 41: Fundación de arte contemporáneo." http://c-41.org/concepcion-41/.

Curiel, Marlin-Stephanie. 2006. "Truth and Consequences: Art in Response to the Truth and Reconciliation Commission." In *Text and Image: Art and the Performance of Memory*, edited by Richard Cándida Smith. New Brunswick: Transaction Publishers.

Duggan, Patrick, and Mick Wallis. 2011. "Trauma and Performance: Maps, Narratives and Folds." *Performance Research* 16 (1): 4–17. https://doi.org/10.1080/13528165.2011.562674.

Fischer, Edward F. 2001. *Cultural Logics and Global Economies: Maya Identity in Thought and Practice*. Austin: University of Texas Press.

Galindo, Regina. n.d. "Regina José Galindo." www.reginajosegalindo.com.

– 2009. "La Conquista." http://www.reginajosegalindo.com/la-conquista/.

– 2012. Interview by David Schmidt. "Entre la Violencia y la Ceguera." *Ágora speed: Postliteraturas* 5/10, no. 1. agoraspeed.org. Accessed May 2015.

– 2014. Interview by Televisionet. "Estoy viva di Regina José Galindo: Al PAC di Milano prende parola il corpo" 25 March. https://www.youtube.com/watch?v=UXUmcOVsHFk.

Galindo, Regina, and Rosa Chávez. 2013. "Inmonumental." 5 January. http://c-41.org/projects/inmonumental/.

Goldman, Francisco. 2006. "Regina José Galindo." Translated by Ezra Fitz and Francisco Goldman. *BOMB Magazine*. https://bombmagazine.org/articles/regina-josé-galindo/.

Hernández, Oswaldo J. 2010. "Latir sin descanso." *Magacín Siglo 21*, 14 November. www.magacin21.com. Accessed 12 January 2015.

Human Rights Watch. 2011. *World Report of 2011: Events of 2010*. New York: Human Rights Watch.

Lavery, Jane, and Sarah Bowskill. 2012. "The Representation of the Female Body in the Multimedia Works of Regina José Galindo." *Bulletin of Latin American Research* 31 (1): 51–64. https://doi.org/10.1111/j.1470-9856.2011.00606.x.

Mejía, Selene. 2015."Regina Galindo se expresa con música en el performance Culpable." *Soy502*, 11 May. http://www.soy502.com/articulo/regina-galindo-expresa-musica-performance-culpable.

Nelson, Diane. 1999. *A Finger in the Wound: Body Politics in Quincentennial Guatemala*. Berkeley: University of California Press.

Olivares, Lissette. 2011. "¿Arte ≠ Política?: In Defense of Performance Praxis." *E-misférica* 8 (1). http://hemisphericinstitute.org/hemi/en/e-misferica-81/olivares.

Organización Internacional para las Migraciones. 2013. *Perfil Migratorio de Guatemala 2012*. Ciudad de Guatemala: Organización Internacional para las Migraciones. https://publications.iom.int/books/perfil-migratorio-de-guatemala-2012.

Peacock, Susan C., and Andrian Beltrán. 2003. *Poderes Ocultos: Grupos ilegales armados en la Guatemala post conflicto y las fuerzas detrás de ellos*. Washington DC: WOLA.

Prensa Comunitaria. 2015. "CULPABLE: Coro para conmemorar los dos años de Sentencia por Genocidio contra Efraín Ríos Montt." Facebook, 10 May. https://www.facebook.com/Comunitaria.Prensa/posts/750295488421626.

Roach, Joseph R. 1996. *Cities of the Dead: Circum-Atlantic Performance*. New York: Columbia University Press.

Schirmer, Jennifer G. 1998. *The Guatemalan Military Project: A Violence Called Democracy*. Philadelphia: University of Pennsylvania Press.

– 2003. "A Violence Called Democracy: The Guatemala of Rios Montt." *ReVista: Harvard Review of Latin America* (Fall): 35–8. https://revista.drclas.harvard.edu/book/violence-called-democracy.

Sileo, Diego. 2014. "Regina Jose Galindo: Necropower." In *Estoy Viva*, edited by Diego Sileo and Eugenio Viola. Milan: Skira Editore S.p.A.

Sommer, Doris. 1991. "Rigoberta's Secrets." *Latin American Perspectives* 18 (3): 32–50. https://doi.org/10.1177/0094582X9101800303.

Taylor, Diana. 2010. *"Save As … Knowledge and Transmission in the Age of Digital Technologies."* *Foreseeable Futures #10.* Imagining America; http://imaginingamerica.org/wp-content/uploads/2015/08/Foreseeable-Futures-10-Taylor.pdf.

– 2013. *The Archive and the Repertoire: Performing Cultural Memory in the Americas.* Durham, NC: Duke University Press.

– 2014. "Introduction: Performance and Politics." *Identities (Yverdon)* 21 (4): 337–43. https://doi.org/10.1080/1070289X.2014.874349.

Tedlock, Dennis. 1996. *Popol Vuh: The Definitive Edition of the Mayan Book of the Dawn of Life and the Glories of Gods and Kings.* New York: Simon and Schuster.

Vela Castañeda, Manolo. 2015 "Guatemala, el circo de la corrupción." *elPeriódico* 28 June. www.elperiodico.com.gt Accessed 30 June 2015.

8   Human and Environmental Justice in the Work of Rodrigo Rey Rosa

STEPHEN HENIGHAN

Introduction

Rodrigo Rey Rosa (b. 1958) is the most widely read novelist of contemporary Guatemala and joins the Salvadoran Horacio Castellanos Moya (b. 1957) and the Nicaraguans Sergio Ramírez (b. 1942) and Gioconda Belli (b. 1948) in the tiny group of Central American novelists with significant international readerships. In spite of the high esteem in which he is held by readers of Spanish, and the publication of his work in other countries, most extensively in the United States and France, the unconventional trajectory of Rey Rosa's career has resulted in his marginalization within the canon of Spanish American literature, particularly his relative neglect in scholarly critical discourse. This chapter traces Rey Rosa's development from a position of aesthetic commitment and relative distance from his homeland's historical debates to a pointed probing of Guatemala's environmental crisis, and an engagement with the injustice inflicted by the civil war years (1961–96), and the pervasive corruption, violence, impunity, and racism against Indigenous people that have characterized the period since the conflict's conclusion. At the same time, this study underlines continuities in Rey Rosa's work, notably his persistence in writing in the non-commercial literary form of the short novel, and his honing of a pared-down, elliptical style that exerts significant demands on the reader to create the meaning of the narrative.

Rey Rosa's anomalous position is underlined by the fact that, though his work is published by the most influential houses in the Spanish-speaking world, first Seix Barral, then Alfaguara, his fiction has been overlooked by scholars in the field, particularly in the Anglo-American academy.

Rey Rosa is the author of twelve short novels, only three of which exceed 150 pages in length, four collections of short stories, a handful of uncollected stories, a travel book about India, and *La cola del dragón* (2014a; The dragon's tail),[1] a book of polemical articles on contemporary Guatemala that appeared – though the appearance may be deceptive – as a rupture from his previous work. The generic murkiness of Rey Rosa's fiction, which employs motifs from science fiction and psychological thrillers, complicates his assessment as a literary artist. The boundaries of his *obra* are opaque: his early short fiction was characterized by Paul Bowles as "often mere scenes or prose-poems of atmosphere, rather than tales" (Bowles [1991] 1997, 119).[2] Though he won the Miguel Ángel Asturias National Literary Prize in Guatemala in 2004 and the prestigious José Donoso Iberoamerican Literary Prize, awarded in Chile, for his life's work, in 2015, Rey Rosa's fiction has been little studied. A central paradox of his career is that his aesthetic inheritance from the U.S. avant-garde, most directly his longstanding mentor Paul Bowles (1910–99), has made him less "teachable" and "researchable" in the academic world by eliding the social themes that are expected of Central American writers, or presenting them in ways that readers find puzzling.

In contrast to writers of the Latin American Boom, who plumbed their nations' histories, Rey Rosa appears diffident about his nationality; he has commented, "For me it's an accident, being Guatemalan" (Gray 2007, 181). The academy's neglect of Rey Rosa's work is illustrated by his omission from Roy C. Boland Osegueda's "The Central American Novel" in Efraín Kristal's highly influential *The Cambridge Companion to the Latin American Novel* (2005). His absence is doubly startling in light of the fact that Boland Osegueda (2005) refers to the work of Guatemalan novelists who are far less widely published, such as Dante Liano (b. 1949), Arturo Arias (b. 1950), and Méndez Vides (b. 1956). Rey Rosa's fiction is not a comfortable fit with Boland Osegueda's contention that "the testimonial impulse continued to beat strongly, particularly in Nicaragua, Guatemala, and El Salvador, in the 1980s and 1990s" (174). Testimonio, defined by John Beverley as a life story "told in the first person by a narrator who is the real protagonist or witness of the events she or he recounts" (Beverley 1993, 70), emerged as an influential genre in the 1970s and 1980s, in reaction to pervasive human rights abuses in Central America and the countries of the Southern Cone, and in frustration at the perceived inability of the morally ambiguous Latin American Boom novel of the 1960s and early 1970s to take sides in the hemisphere's ideological struggles.[3] The testimonial impulse extended to fiction, engendering works that absorbed the voice and conventions of testimonio into the novel.[4] As a result, Anglo-American criticism, as exemplified by Boland Osegueda, often defined Central American literature in terms of testimonio. Among the handful of scholarly articles on Rey Rosa's fiction, studies of *El material humano* (2009; Human material), the sole work of fiction in which he adopts an incipiently testimonio voice, are over-represented.[5] Most of his novels, though highly praised, have not been studied. In the most critically sophisticated book-length study of Central American literature to date, *Taking Their Word: Literature and the Signs of Central America* (2007), by his compatriot Arturo Arias, Rey Rosa appears precisely once, in a long list of the names of writers who are "producing fascinating new postwar novels" (23).

A key issue here is exoticization, both in terms of Rey Rosa's restless travels and in the ways in which he filters Guatemalan reality through Paul Bowles's quest for locales in which to escape the monotony of Western modernity. Born into a wealthy family that was Guatemalan mestizo on his mother's side and Italian on his father's side, Rey Rosa was educated at the Jesuit Liceo Javier in Guatemala City. In the late 1970s and early 1980s, he followed other young Guatemalans who had the means to do so into voluntary exile. Yet his destination was incongruous. Central Americans who were active in the political opposition congregated in Mexico City.[6] Making a decision consistent with his social class, Rey Rosa chose to study cinema at the School of Visual Arts in New York City. He dropped out of film school, worked at odd jobs, and travelled in Europe and North Africa, spending most of the years between his departure from Guatemala in 1978 and his definitive return to the country in 1994 in Tangier, New York, Barcelona, or Paris. Even after returning to Guatemala, he continued spending summers in Tangier until 2001 (Goldman 2013). Rey Rosa's wandering, and his association with a legendary figure of the U.S. avant-garde, made him, in turn, a legend among avant-garde Spanish American writers of his generation.[7]

Rey Rosa's Moroccan-US literary apprenticeship, compounding his upbringing in the Guatemala City upper classes that live at a remove from national reality, and the conditions of the war when, as Rey Rosa notes, the violence in the countryside was suppressed from the newspapers – "era practicamente imposible para el ciudadano medio enterarse de lo que ocurría, porque la prensa callaba casi todo" (Albacete 2014; it was practically impossible for the average citizen to learn what was going on because the press was silent about it all) – layered on top of his father's Italian immigrant roots, created a literary dynamic where the author approached Guatemala as a country to be understood from the outside, through travel and observation, as he might understand India, Morocco, or Mali, rather than written about from the gut, as his near-contemporary, regional counterpart, and friend Horacio Castellanos Moya does in his early novels about El Salvador, most flagrantly in *El asco* (1997; *Revulsion*, 2016).[8] Rey Rosa's exile was the essential preparation for his re-imagination, after 1994, of Guatemala's geography, threatened physical environment, and history of racism and social injustice.

Influences: Paul Bowles, Jorge Luis Borges

The prolific writer, musical composer, and translator Paul Bowles settled in Tangier in 1947. Bowles, who incarnated the American abroad in its most extreme personification, was the successor to the Lost Generation of Ernest Hemingway, F. Scott Fitzgerald, and Gertrude Stein in the Paris of the 1920s. Though much romanticized, the Lost Generation was an ingrown expatriate nucleus that ignored French society, politics, and artistic currents. From the 1930s onwards, escape from the commercialized drabness of the United States required a willingness to engage with the culture of the country of expatriation. As Martha Gellhorn noted of her arrival in Paris in February 1930, "Unlike the gifted Americans and British who settled in Paris in the Twenties and lived in what seems to me a cosy literary world, I soon lived entirely among the French, not a cosy world" (Gellhorn 1989, 91). Bowles, who made a short-lived attempt to join the Paris expatriates at the age of nineteen (Halpern 2002, 907), was the human link between the Lost Generation and the Beat culture of the post–Second World War era. Beat writers rebelled against a triumphant, imperial, economically successful, yet increasingly regimented US society by immersing themselves in cultures that they viewed as more raw, vital, and authentic. The Beats expressed their revolt through perpetual wandering, drug use, and acts of sexual non-conformism, laying the foundations for the more widespread 1960s youth culture of soft drugs and sexual freedom. Once Bowles and his wife, the writer Jane Bowles (1917–73), both of whom were bisexual and took same-sex Moroccan lovers, were established in Tangier, the city became a place of pilgrimage for US literary rebels.

Paul Bowles was a minor avant-garde writer until the publication of *The Sheltering Sky* ([1949] 2002), one of the most accomplished US novels of the mid-twentieth century. *The Sheltering Sky* distilled the tensions in Bowles's personal life and his desire to cross over into alien cultures. Kit, the female protagonist, complains, "The people of each country get more like the people of every other country. They have no character, no beauty, no ideals, no culture – nothing, nothing" (Bowles 2002, 8). In reaction against the homogenizing forces of post–Second World War capitalism, Port and Kit, their marriage in crisis after infidelities by both partners, make a journey into the deserts of northwest Africa. Port falls ill and dies; the narrative consciousness passes to Kit, who wanders by herself through the desert, is picked up by a camel train, sexually used by Arab men, and becomes a trader's concubine. She escapes and is repatriated by diplomats, apparently to her deep regret: the novel concludes with an image of a streetcar, epitomizing the mechanized, emotionally vacant world to which Kit has returned. Kit lives out in symbolic form the physical possession by the primitive that the male protagonist cannot enjoy. Even though her experience is not sustainable, it incarnates an ideal of escape from soulless modernity.[9]

During the tentative early stages of the Bowles revival that would peak after 1990, when Bernardo Bertolucci's film of *The Sheltering Sky* (1990) opened with shots of Bowles, now nearly eighty years old, sitting in a café in Tangier, Rodrigo Rey Rosa arrived in Morocco as a student in the 1980 summer semester abroad offered by the School of Visual Arts (Carr 2004, 305). Initially more interested in cinema than in literature, Rey Rosa had begun to write short stories during the period between his first solo trip around Europe at eighteen and his departure for New York. His desire to write was sparked by reading the short stories of the great Argentine fabulist Jorge Luis Borges (1899–1986). Rey Rosa recalls, "Back in Guatemala, reading Borges – I remember exactly, reading *Ficciones*, I decided I wanted to really, you know, try to do that.… To create worlds like that. That was my ambition, rather than being read by millions" (Gray 2007, 185).[10]

In order to qualify for admission to Bowles's creative writing workshop, Rey Rosa wrote a few pages in English. Bowles recalled, "He spoke very little English, and like most of the other students, had never submitted anything for publication.[11] I don't know how he thought he could profit from the class. No one else spoke Spanish, but he knew intuitively that I did, so we communicated that way, and everything he wrote was in Spanish. He had a great imagination and knew how to create atmosphere quickly" (Carr 2004, 305).

In the first class, Bowles asked the students their places of birth and their favourite writers. After the class, Bowles approached Rey Rosa, "para decirme en español que él había viajado por Guatemala y México, y que si el inglés no era mi lengua materna, que escribiera en español, que él no tenía dificultad para leerlo. Borges era también un autor de su predilección, agregó, y lo leía en español" (Rey Rosa 2014a, 22; "to tell me in Spanish that he had travelled through Guatemala and Mexico, and that if English wasn't my mother tongue, I should write in Spanish, that he read it without difficulty. Borges was also one of his favourite writers, he added, and he read him in Spanish").[12]

This was the beginning of an intense professional relationship. Rey Rosa wrote his short stories with Bowles in mind as his ideal reader: "Paul was my audience for a long time" (Gray 2007, 172); the Guatemalan, in turn, translated Bowles's work into Spanish. Rey Rosa dropped out of the School of Visual Arts and spent most of 1984 writing in Morocco: "My social life was limited to Bowles's tea hour. I would go there and have tea, and I met many, many people there" (175). In a feature article published in *Vanity Fair* magazine in 1985, the novelist Jay McInerny describes these gatherings:

[Bowles] rises to greet his visitors, of whom there is a steady influx, extending a hand and addressing them in whatever language is appropriate – French, Spanish, Arabic, or English. He is extremely polite and courtly in manner.... The conversation – about Tangier, about music and writing – proceeds in English and Spanish. Rodrigo Rey Rosa, a young Guatemalan writer whose work Bowles is translating, arrives with his girlfriend.... Asked if he particularly admires any living American writers, Bowles ... looks earnestly at the ceiling.... When the field is expanded to include South America, he is able to endorse Jorge Luis Borges without hesitation. (Caponi 1993, 182–3)

The collaboration between Rey Rosa and Bowles produced three books translated by Bowles and published in English.[13] Rey Rosa was not a natural member of Bowles's inner circle, most of whom were homosexual men; as he said, "We didn't have much in common" (Gray 2007, 166). Bowles's letters to his young protegé, written in Spanish, address Rey Rosa with the formal *usted* (Bowles 1994, 504–5). Yet by November 1990, the old man's delight in the younger writer's presence is palpable in a letter to American photographer Cherie Nutting: "Rodrigo comes every afternoon at tea time. How he manages to walk all the way from Charif[14] in the pouring rain, I don't know. He has typed the entire book on Mali, for which I'm most grateful" (544). The old American and the young Guatemalan were drawn together also by Bowles's fascination with Mexico and Central America.[15] Imagining Bowles as his audience, Rey Rosa absorbed a vision of the isthmus as construed by the U.S. expatriate avant-garde. A common self-definition as expatriates and a shared interest in travel writing evolved into a congruent approach to the depiction of Central America. Rey Rosa recalls, "I became particularly interested in English-language writers who had written about Latin America, about Mesoamerica in particular; I suppose they showed me a Latin American, or Central American, countryside that is almost impossible to find in Hispanic literature. ([Juan] Rulfo is the exception here)" (Esposito 2015). Bowles's novel *Up above the World*, about an American couple whose lives become entangled with those of a group of upper-class Central Americans, appears to have served as a rough template for Rey Rosa's own depictions of the isthmus. The outbursts of irrational cruelty by characters who are initially sympathetic, which in Bowles short stories such as "At Paso Rojo" (Bowles 2001, 60–76) stem from the American's perception of foreign cultures, particularly in Mexico and Central America, as dangerous and different, are internalized by Rey Rosa. Cruelty enters the Guatemalan's repertoire for understanding his own country as a literary mannerism; but, after Rey Rosa's return to Guatemala, cruelty and violence become anchored in concrete historical circumstances, undergoing a gradual transformation from aesthetic trait to social criticism. Bowles's aesthetics of clear, precise language and narrative restraint induce the reader to enter into the action, dispensing with the explanations of an interventionist narrator. This aesthetic became an enduring influence in Rey Rosa's work, as did the Beat movement's idealization of nonconformism.[16] Though submerged in the early years, Beat-inspired nonconformism would propel Rey Rosa's portraits of Guatemalan society.

The Career Path Doubles Back: Paradoxes of Literary Identity

In approaching his national reality through the lens of a metropolitan avant-garde movement, Rey Rosa recapitulated the trajectory of the Nobel Prize–winning Guatemalan novelist Miguel Ángel Asturias (1899–1974), who forged his notion of Guatemalan identity in response to the Indigenous-oriented expectations of French surrealist writers whom he met in the 1920s (Henighan 1999, 28–79). The crucial difference between these two major Guatemalan writers of fiction, born sixty years apart, is that Asturias, who defined himself in Paris as the voice of the Mayan people (though, like Rey Rosa, he came from an well-off mestizo family in Guatemala City), did not engage more deeply with Guatemala, in particular rural Guatemala, in later life.[17] Rey Rosa, in contrast to Asturias, did not assume a Mayan identification; his national focus embraced Guatemala's ethnic diversity and recognized the ethnic mix of his own background. Bowles wrote, "He is very conscious of the Mayan past of his country without being obsessed by it" (Bowles [1991] 1997, 119–20). The autobiographical narrator of *El material humano* tells an acquaintance, "Tenemos algo de mayas, pero nuestros nombres son europeos, y tenemos sangre italiana por parte de padre. Pero también somos descendientes de los conquistadores. ¡Somos también los malos!" (Rey Rosa 2009, 130; We have some Mayan ancestry, but our names are European and we have Italian blood on the paternal side. But we're also descended from the conquistadors. We're the bad guys, too!). After his return to the country in 1994, Rey Rosa continued to travel in other countries; yet, in contrast to Asturias, he travelled throughout his country and immersed himself in Guatemalan realities beyond those of his own social class. Over time, Rey Rosa's depictions of Guatemala grew more complex, troubling, and politicized.

This later trajectory was belied by his position outside, and practically in opposition to, the dominant tendencies of Spanish American literature. Bowles's mentorship, and their common enthusiasm for Borges, were both liabilities. The U.S. literary influence sanctioned by the writers of the Boom was William Faulkner (1897–1962). Gabriel García Márquez (1927–2014), whose *Cien años de soledad* (1967; *One Hundred Years of Solitude*, 1970) was in some ways a rewriting of Faulkner's modernist tale of the decline and fall of a Southern family in *Absalom, Absalom!* ([1936] 1964), and Mario Vargas Llosa (b. 1936), whose major novel *La casa verde* (1965; *The Green House*) was a rewriting of *The Wild Palms* ([1939] 1962), had enshrined Faulkner as the U.S. touchstone for Spanish American writers. Faulkner was a permissible Yanqui influence, even at the height of Latin American intellectuals' enthusiasm for the Cuban Revolution in the 1960s, because he came from the region of the United States whose military defeat was the precondition for the creation of the expansionist United States of the late nineteenth and early twentieth centuries that imposed its "Big Stick" policy on the

Caribbean and Central America (LaFeber 1983, 19–83). By contrast, the decadent apoliticism of the homosexual, drug-taking Beats found no echo in a continent whose artists and intellectuals engaged in masculinist posturing to bolster national self-assertion and social equality amid the ideological strife of the Cold War. The Boom legacy of complex, historically engaged novels made minimalist short stories that dramatized violence as an aesthetic experience – one possible reading of *The Beggar's Knife* – seem un-Spanish American. Borges may have been the Boom novel's indispensable ancestor, opening up a cosmopolitan dialogue with world literature, but he was an ancestor whose legacy had been tainted. As García Márquez's biographer states, "García Márquez takes up many of Borges's ideas (though without acknowledging this new influence)" (Martin 2008, 144). In the ideologically divided early 1980s, with civil wars raging in three Central American countries (Nicaragua, El Salvador, Guatemala), the U.S. military in de facto occupation of a fourth country (Honduras), the Dirty War in full swing in Argentina, and Chile, Uruguay, and Brazil ruled by U.S.-installed dictatorships, Borges sided with the tyrants. In 1976 he ate a widely publicized lunch with General Jorge Videla, the architect of policies under which 30,000 Argentine civilians were "disappeared." Borges later accepted the Grand Cross of the Order of Merit from Chile's General Pinochet, a miscalculation that cost him the Nobel Prize for Literature (Williamson 2004, 424–5). By the early 1980s Borges had become mildly critical of the generals, though he never reprimanded them without at the same time criticizing the left far more harshly (451–5). Rey Rosa, opposed to the generals who ran his country, never travelled without "las *Obras completas* de Borges en la edición EMECÉ, que yo frecuentaba como supongo que un alma religiosa frecuentaba la Biblia (ningún día sin leer por lo menos un poema de aquel grueso libro de tapas verdes)" (Rey Rosa 2014, 101; "the *Complete Works* of Borges in the EMECÉ edition, which I consulted as I suppose a religious soul would consult the Bible [letting no day pass without at least a poem from that thick book with green covers]"). It was an unpropitious historical moment in which to be a follower of Borges. Compounding these obstacles was the fact that Rey Rosa, who did not work as a journalist or editor, or teach at a university, lived outside the ambit of Latin American cultural institutions.

With *The Path Doubles Back* (a chapbook translation of "El camino se dobla" published in New York in 1982), *The Beggar's Knife, Dust on Her Tongue,* and *The Pelcari Project,* Rey Rosa was first published in English in New York, San Francisco, and London. Uniquely among Spanish

American writers published in English, Rey Rosa was not read internationally in Spanish. By being published in English, while remaining unknown in his own language, Rey Rosa's work appeared, at the height of the Cold War, as a figment of U.S. decadentism's delusions about Latin America. He published his early books in war-torn Guatemala, where almost no one read, and the bourgeoisie, radicalized to the far right, was actively hostile to literature.[18] Publishing in Guatemala confined Rey Rosa to being an exclusively Guatemalan writer in Spanish, in spite of being an international writer in English. His breakthrough occurred when the Catalan poet Pere Gimferrer, an editor for Seix Barral, read a review of Rey Rosa's work in the *Times Literary Supplement*: "Seix Barral contacted me. I never dared send anything … Seix Barral asked, and I sent the books" (Gray 2007, 179–80).

Rey Rosa's association with Seix Barral leads to another paradox: his internationalization in Spanish coincided with his return to Guatemala. In 1992 Seix Barral published an international edition of *The Pelcari Project* and a new Borgesian novella, *El Salvador de Buques* (The saviour of ships) as a single volume; in 1994 *Lo que soñó Sebastián* appeared shortly after Rey Rosa had re-established himself in Guatemala. The influence of Borges, and later of Adolfo Bioy Casares (1914–99), who collaborated with Borges for more than fifty years, inducted Rey Rosa's fiction into the neo-fantastic genre. The critic Jaime Alazraki has defined this mode as the recapitulation, in a Spanish American context, of the conditions that created the short story of the fantastic at the beginning of the nineteenth century in early-industrializing regions such as Germany and the East Coast of the United States, producing writers such as E.T.A. Hoffman (1776–1822) and Edgar Allan Poe (1809–49), whose work exploited the newly urbanized population's loss of contact with the natural world to generate fear through the evocation of phenomena that appear to defy rational logic. Alazraki sees a similar development, which he refers to as "lo neofantástico" (1990), fuelling the emergence of fantastic short stories in mid-twentieth-century Buenos Aires. Alazraki's definition inspired a critical shorthand in which the neo-fantastic became the anti-realist literary mode of Buenos Aires and Montevideo, while in less urbanized regions the cosmovisions of animist cultures, either Indigenous or African-descended, were incorporated into fiction through adaptation of modernist techniques to transculturate national histories (Rama 1982, 107–9). This latter tendency produced the various forms of the marvellous real and magic realism practised by writers such as Alejo Carpentier (1904–80), Miguel Ángel Asturias,

José María Arguedas (1911–69), and Gabriel García Márquez. According to this shorthand, writers from less urbanized areas, with stronger Indigenous presences, would write magic realism. Reversing this paradigm, Rey Rosa immersed himself in the neo-fantastic tradition. The fact that Guatemala had already, in Asturias, produced a major novelist who transculturated national realities through quasi–magic realist techniques may have made innoculated Rey Rosa (whose most obvious Guatemalan precursor, in aesthetic terms, is Augusto Monterroso) against this approach. After his return to Guatemala, Rey Rosa wrote fewer short stories and more short novels. While remaining loyal to the concision of Borges and Bowles, Rey Rosa was seeking ways to expand his narrative frames in order to dramatize more complex social interactions. Borges, who never wrote a novel, offered no model for expanding neo-fantastic aesthetics to novel length; yet a model for a neo-fantastic novel was supplied by Adolfo Bioy Casares.

The cerebral themes of Borges – the world as a tissue of cross-references, recurrent symbols whose eternal return nullifies linear time, the isolation of the male intellectual, reading as an adventure – offered both a path into literature and an impediment to expanding a work of fiction beyond the brevity of a short story. Bioy Casares had found a solution to this dilemma in novels such as *La invención de Morel* ([1940] 1982; *The Invention of Morel*), *Plan de Evasión* ([1945] 1969; *A Plan for Escape*) and *Dormir al sol* (1973; *Asleep in the Sun*), where he built Borgesian themes around a neo-fantastic object or project – a scientific or technological breakthrough, for example, that involved human experimentation – in order to open up spaces for character development and a novel-length plot. Rey Rosa comments, "As for Bioy – whom I continue to read – his work helps me somehow to bridge the gaps, to withstand Borges's influence" (Esposito 2015). Rey Rosa's first two attempts at long fiction, *The Pelcari Project* and *El Salvador de Buques*, both follow the model of Bioy's novels. *The Pelcari Project* is strongly reminiscent of *The Invention of Morel*, whose narrative voice it mimics perfectly; the name Pelcari is drawn from a reference in the final pages of Bioy's *A Plan for Escape*. Like Bioy's early novels, Rey Rosa's are "islands of male conversation" (Henighan 2012, 94–101), absorbing the male exclusivity of both the Borges-Bioy friendship[19] and Bowles's inner circle. The only female character of any significance is Dr Pelcari herself, who appears briefly in the novel's opening and closing pages. *The Invention of Morel* and *A Plan for Escape* also supplied the structural models for Rey Rosa's longer fiction. He would continue to exploit Bioy-like sinister uses of technology – a reflection, also, of Bowles's Beat-era suspicion of the dehumanizing propensities of modernity – to furnish his novels with plots: *The Pelcari Project, El Salvador de Buques, Que me maten si …* (1997; Let them try to kill me …), *Caballeriza* (2006; Stud stable), and *Los sordos* (2012; Deaf people) all depend on conspiracies and nefarious inventions to set the action in motion. In the novels in which Rey Rosa's vision is most distinctive, particularly *Que me maten si …* and *Los sordos*, it is the tension between the attempt to enfold swathes of Guatemalan geography with extreme succinctness, paradoxically linked to incipiently science fiction–like technology or human experimentation, that lends his work its inimitable flavour. The novels that do not employ such devices are either autobiographical and/ or set in Guatemala City. *El cojo bueno* (1996; *The Good Cripple*) is a form of altered autobiography, *La orilla africana* (1999; *The African Shore*), published in the year of Paul Bowles's death, is spiritually autobiographical, in the sense that it is "a double farewell" (Esposito 2015) to both his mentor and the city of Tangier; in *El material humano*, the line between author and narrator becomes indiscernible. This latter novel is dominated by the society and geography of Guatemala City; this is also true of *Piedras encantadas* (2001; Enchanted stones), and, though less explicitly so, of *Severina* (2011), where a Borgesian gambit – the narrator owns a bookstore – evolves into a plot reminiscent of the Tom Ripley novels of Patricia Highsmith (1921–95), in which a sophisticated man driven by powerful desires finds himself committing murder and hiding a body.[20]

The geography of Rey Rosa's early fiction is only vaguely Guatemalan. The location of the prison camp in which the central character of *The Pelcari Project* is confined is decipherable. The jungle hospital he reaches after escaping from the prison is in Belize (*Cárcel*, 17; *Pelcari*, 27). At the novel's conclusion, Dr William Adie, who must try to solve the riddle of whether the bedraggled figure before him can be the same man who wrote the erudite journal found in his possession, which chronicles his efforts to survive after undergoing experimental brain surgery, notes that the man speaks Spanish, and that the countries to the north and west are both Spanish speaking. He decides that this man probably "venía del río arriba, del Oeste" (*Cárcel*, 39; "the man came from up-river, from the west," *Pelcari*, 93). The country to the west of Belize is, of course, Guatemala. Whether this extreme discretion in signalling his own country as a place of repression is a product of the intimidation of the war years, or, more likely, an absorption of the reluctance of Borges and Bioy Casares to provide clear solutions to their puzzles, it betrays a tension in Rey Rosa's early work between the urge to remain aloof to Guatemalan injustice and the need to speak about it. *The Pelcari Project* articulates its critique of oppression in abstract terms; Guatemala's colonial history and ethnic divisions are displaced by chapter headings adorned with epigraphs from Borges and Ludwig Wittgenstein. The prisoners whose heads are covered in scars from experimental brain surgery belong to different racial groups; when the narrator reaches Belize at the outset, he is disparaged as a "nigger" (*Cárcel*, 16; *Pelcari*, 24). Yet the existence of racial diversity, and racial prejudice, is not contextualized. Rey Rosa's critique of authoritarianism dovetails with his metaphysical concern with male isolation inherited from Borges and Bioy Casares: the solution to authoritarianism is dialogue. The narrator's oppression recedes when he begins to share his notebook with a fellow prisoner, whom he knows only as Yu. The two men overcome alienation and ultimately imprisonment itself by writing alternating entries; in a reprise of the themes of Bioy Casares's early novels, male conversation breaks down isolation.

Rey Rosa's second short novel, *El Salvador de Buques*, is even more claustrophobic and rarefied. Set in a neo-Borgesian world of libraries, private studies, and conference rooms, in a location that shares discreet geographical references with Central America, in a vein similar to *The Pelcari Project*, the work narrates the struggle of the deranged "admiral" to defeat a conspiracy led by Dr Fernández. The narrative universe is confined to people in positions of power. Those who do not belong to the upper echelons of society are absent from this work, which acknowledges no social, cultural, or racial differences or injustices. Rey Rosa's return to Guatemala would make this sort of fiction unsustainable.

Environmental Injustice: *Lo que soñó Sebastián*

Rey Rosa's return to Guatemala challenged him to adapt his art of concision, suggestion, and indirection to circumstances in which it was impossible to exclude the subject matter of the dominant Central American literary modes of testimonio narrative and politically committed poetry: the political oppression, grotesque injustices, and horrendous violence that characterized recent decades of history. The tension between these imperatives created much of his best work. Had Rey Rosa remained in Morocco after Bowles's death, or settled in Barcelona, as Bolaño did in his later years, or gone to Paris, where his friend the Catalan artist Miquel Barceló lived, he might have developed into a figure roughly analogous to that of V.S. Naipaul (b. 1932) in English literature, brandishing purity of style as a banner of defiance to the anti-imperialism of other writers from less-developed countries who remained more connected to their homelands, and roaming remote corners of the world as the ghost of Paul Bowles, much as Naipaul travelled as the ghost of Joseph Conrad. Rey Rosa's return to Guatemala forced him to modify his identification with his literary influences. Inevitably, in a Hispanic society, his return home was a return to his family: a father who was an industrialist, a mother who had survived a kidnapping, two sisters, one of whom, Magalí Rey Rosa, was the country's most prominent environmentalist – "tachada de 'ecoterrorista' por una serie de columnas de prensa" (Rey Rosa 2009, 82–3; accused of being an "eco-terrorist" by a succession of newspaper columns) – various uncles and aunts and nephews and nieces, and, after 2002, a daughter, whose presence in Guatemala City Rey Rosa cited as one of his reasons for remaining in the country (Goldman 2013). The daily realities with which Rey Rosa came into contact as a member of a Guatemalan family nourished his fiction, heightening what Harold Bloom would term the "anxiety of influence" with regard to his fealty to Borges. Though Bloom based his theory on English poetry rather than Spanish American prose, its application to Rey Rosa's fiction is apt, because Bloom's notion that young writers deliberately misread their most influential predecessors as part of the process of developing a recognizably individual style is derived – this is Bloom's own anxiety of influence – from Borges's theory that each writer creates his precursors by recasting Miguel de Cervantes or Franz Kafka as the writer this illustrious predecessor must become in order to nourish the younger writer's creative quest (Bloom 1973, 19–45). While Bloom carries his theorizing to whimsical extremes that are not germane to the present analysis, his insights assist in understanding how the author of *The Pelcari Project* and *El Salvador de Buques* became, a few months later, the author of *Lo que soñó Sebastián*. Rey Rosa's statement that the fiction of Bioy Casares enabled him "to withstand Borges's influence" (Esposito 2015) alerts the reader to his strategy of rewriting one influence through another. In Bloomian (and also Borgesian) terms, Rey Rosa rewrites Borges as Bioy Casares in order to gain access to the form of the novel; he then rewrites Bioy Casares as Paul Bowles in order to engage with the Central American landscape. This formulation is too neat, of course; but it is broadly accurate, as the novel's precisely rendered descriptions of the jungle, wildlife, and people who inhabit remote lagoons and rivers attest.

Rey Rosa's primary concern in *Lo que soñó Sebastián* is environmental. He re-enters Guatemala, like Paul Bowles venturing into an unknown region of North Africa, through the country's landscape. Before he can address questions of social justice or historical clarification, he must define the physical terrain. Two of the first three novels he wrote after his return to Guatemala, *Lo que soñó Sebastián* and *Que me maten si ...*, undertake this task in explicit form. His initial concern for the country – "El país más hermoso, la gente más fea" (Rey Rosa 2001, 9; the most beautiful country, the ugliest people) as he would later write of the Guatemala City upper classes – takes the form of seeking to preserve its abundant natural life. Environmental justice is first cast as conservation; then the novel problematizes this assumption.

The Petén region of northeastern Guatemala is the largest freshwater wetland area in the Western Hemisphere north of the Amazon Basin; in recent years, large sections have been burned to "open up" the land for cattle ranching or have been turned over to foreign extractive industries, as described by Kalowatie Deonandan and Rebecca Tatham in chapter 3 and Magalí Rey Rosa in chapter 4. As Rey Rosa wrote in his 2014 introduction to two jungle stories that appeared at the same time as *Lo que soñó Sebastián*, "Buena parte de la selva que registran ha sido convertida en tierra arada o en territorio narco" (Rey Rosa 2014, 12; A large part of the jungle they record has been converted into tilled land or narco territory). *Lo que soñó Sebastián*, which foresees this environmental catastrophe, dramatizes the counter-productive consequences of a solitary bourgeois idealist's decision to buy a house in the jungle and ban the local people from hunting on his land in order to conserve the animals. In a very compact novel of barely 100 pages that swarms with ironies, the hopelessness conveyed by the action is to some degree negated by Rey Rosa's meticulous descriptions of flora and fauna, which leave the reader with an enduring impression of the natural richness that is at stake in the Petén. The intricate, often imperceptible interconnections that sustain ecosystems mimic the tangled ecology of the local society that Sebastián Sosa (whose alliterative name rhymes with that of his creator) disrupts with his insensitive attempt to impose his ideals on a world he does not understand. This point, never explained to the reader by the novel's diaphanous narrative voice, is implicit in Rey Rosa's descriptions of the natural world when Sebastián takes a walk in the jungle with Véronique, a friend who is visiting from France:

A menudo Véronique se detenía para admirar un árbol gigante, un cedro rojo cargado de lianas que subían y bajaban serpenteando, ya buscando la luz, ya huyendo de ella, o un pucté moribundo, rebozante de orugas y de orquídeas, o un canela asfixiado por un matapalo y que, tumbado entre sus compañeros que parecían sostenerlo con sus tentáculos para que no tocase el suelo, era como un andrajosa, desmedidad pietà. (Rey Rosa [1994] 2000, 82)

Often Véronique would stop to admire a gigantic tree, a red cedar loaded with creepers that wound climbing and descending, now seeking the light, now fleeing it, or a moribund black olive tree, coated in caterpillars and orchids, or a cinnamon throttled by a strangler fig tree and which, having fallen over among its comrades, who seemed to be holding it up with their tentacles in order that it not touch the ground, was like a ragged, oversized *pietà*.

With the exception of the incongruous closing allusion to Italian art, which brings a European lens to bear on the Central American natural world, this description employs a Bowlesian precision to suggest an interconnectedness and mutual dependence that is too complicated to benefit from intervention from an uninformed outsider. The creepers that sometimes seek the light and sometimes flee it, and the complicated relationship of the fallen cinnamon tree, which appears to be dead, yet is held off the ground by the strangler figs' creepers, portray the natural world as an impenetrable net of often contradictory interrelationships. By attempting to ban hunting, Sebastián oversimplifies these interconnections. The calamitous consequences of this decision become evident in the opening pages when Roberto, the handsome son of the Cajals, a local hunting family, shoots dead Juventino, whom Sebastián has hired to help him manage his land. The scene opens with Juventino explaining basic ecology to Sebastián, trying to persuade him to contradict his liberal impulses in dealing with plant and animal life. Juventino argues that Sebastián's interdiction on hunting will disrupt migration patterns since the Cajals will cut off the animals' access to the water in order to prevent them from reaching his sanctuary: "Pero si les cortan el paso aquí, los animales nunca llegarán a donde yo puedo esperarlos. Vienen aquí porque aquí se pasa al agua" (Rey Rosa [1994] 2000, 13; But if they block their route here, the animals will never get to a point where I can wait for them. They come here because this is how they get to the water).

If Sebastián ruptures the traditions that maintain an equilibrium among the jungle's life forms, so does Roberto Cajal. In contrast to his uncle, Francisco, who displays a sensitivity to the rhythms of the natural world, for the nephew hunting is both a means of improving his economic status and an obsession. As he complains to Véronique,

> Mientras haya animales en la selva, yo seré cazador. Este señor [Sebastián] no entiende. Él debería ponerse a pelear con los finqueros, que son los que arrasan a los árboles para hacer sus potreros, o con los militares, que por achicharrar a dos subversivos queman no sé cuántas caballerías de selva. Pero en vez de eso se mete con nosotros, que queremos a los animales tal vez más que él. (Rey Rosa [1994] 2000, 76–7)

> As long as there are animals in the jungle, I'll be a hunter. This gentleman [Sebastián] doesn't understand. He should start fighting the farmers, who flatten the trees to make their cattle ranches, or the army who burn I-don't-know-how-many acres of jungle to scorch a couple of subversives. But instead of doing that, he turns on us, who may love the animals more than he does.

Here social stratification impinges on the environmental debate. The Cajals are poor and have a Mayan surname; Sebastián is wealthy, with a Spanish surname and a lawyer in the capital. The American, Richard Howard, who represents a level of globalized privilege that surpasses Sebastián's privileged status within the nation, builds a mansion and high-end tourist lodge on his property on the opposite shore of the lagoon. He hosts a team of archeologists who are excavating a Mayan site. This solves his poaching problem, as the presence of privileged outsiders with security guards scares off the Cajals. The artisan Don Félix tells Sebastián "Lo que le hace falta a tu terreno es una tumba maya … nadie caza en la tierra de Howard" (Rey Rosa [1994] 2000, 72; What you need on your land is a Mayan tomb … nobody hunts on Howard's land). When over-hunting reduces the supply of animals, Francisco Cajal knows he must let the jungle lie fallow, "para dejar pasar la mala racha. Roberto, en cambio, pasaba a solas en la selva mucho más tiempo que antes" (88; to sit out the rough patch. Roberto, on the other hand, spent much more time alone in the jungle than before). Roberto's failure to absorb his uncle's wisdom signals the growing imbalance in traditional structures that threatens the intricate interconnections in the region's ecology. This instability is accentuated by Roberto's efforts to seduce Sebastián's French visitor. Véronique, who sometimes sleeps at Sebastián's house, is not his lover, though an uncomfortable sexual tension exists between them. Sebastián, too, violates lines of culture and social class by having an affair with María Cajal, Francisco's daughter; this makes him unpopular in the local community.

Lo que soñó Sebastián is the first of Rey Rosa's books to portray hetero-sexual desire. His first three books quash this subject. Like Bioy Casares, who, absorbing the patterns of the sexually repressed Borges, concentrated in his early fiction on civilized male conversation,[21] Rey Rosa lacked a literary model for writing about heterosexuality. The troubled couples of Bowles's Central American novel, *Up above the World*, are the closest approximation available to him among the works of his mentor or models. The attempt to portray Guatemalan society necessitates a move towards the depiction of male–female relationships, just as it ushers the Mayan heritage into his work. While Mayan characters or constructions make tangential appearances in early short stories such as, "La lluvia y otros niños" (The rain and other children) in *The Beggar's Knife*, "La prueba" (The proof) and "Xquic" in *Dust on Her Tongue*, and "El cerro" (The hill), which was published in the original edition of *Lo que soñó Sebastián* but omitted from *1986. Cuentos completos*, in *Lo que soñó Sebastián*, the Maya are no longer unknowable. Yet they are relegated to the past. Subaltern characters such as the Cajals, or the police officer Constable Ba, have Mayan surnames (Ba's commanding officer, Godoy, has a Spanish surname). Mayan ruins serve as a pretext for intervention in the ongoing land dispute; travellers through the jungle navigate via an old Mayan canal; there is a discussion, in the past tense, of how the Maya made condoms (Rey Rosa [1994] 2000, 33). The Mayan presence enters Rey Rosa's fiction as an element of national heritage. Unlike the natural world, the Maya cannot be preserved because, as portrayed here, they have already been assimilated. The vestigial Mayan culture is engulfed by corruption in the form of an illicit trade in real and counterfeit Mayan artefacts.

The image that crystallizes the wildlife conservation debate is the margay cub that Roberto Cajal gives Véronique as a gift.[22] Véronique's determination to take the margay back to Paris provokes a crisis between her and Sebastián. When their boat founders as he returns her to the Howards' hotel, she begs him to rescue the margay. "Si no insistes en llevártelo," he replies (Rey Rosa [1994] 2000, 86; If you don't insist on taking him with you). With restrained indirection, Rey Rosa establishes Véronique's naive tourist notion of "saving nature" by taking a margay to Paris as an amplification of Sebastián's naivety in imposing an unenforceable ban on hunting and going out at 2:30 a.m. to make noise to scare the animals out of the hunters' reach. Véronique's naivety makes him aware of the futility of his own acts. Sebastián's decision to release the margay into the jungle after her departure represents his relinquishment of his "dream" that nature's interconnections can survive untouched by human society. The inevitability of the jungle's annexation by an avaricious modernity subsumes Sebastián and Roberto, who are divided by conflicts of social class, and disputes over women and the treatment of animals, into a common institutionalized framework of a series of enclosures. Sebastián sets up Roberto to be arrested by the police for trafficking in Mayan artefacts. Yet, though temporarily successful in eliminating his rival, he sparks greater interest in his property and is asked to approve a licence for archaeological excavations. His final line of dialogue – "Vengan a verme cuando quieran" (100; Come and see me whenever you feel like it) – foreshadows his irrelevance now that institutional structures determine the destiny of his land. In a similar vein, Roberto, once he is released from jail, finds that his uncle, Francisco, has given up hunting and created a series of enclosures in which to raise lowland pacas – large tropical rodents – as meat for the national market. Roberto swears to continue hunting, yet like Sebastián he is impotent to prevent his demotion. The novel's deceptively casual final line is Roberto's observation of his uncle: "vio que había envejecido muchos años" (102; he saw that he had aged many years). The diminishment of the jungle to a shrinking, managed environment of agricultural and primary resource productivity contained within closed-off spaces deprives the Cajals of the freedom to roam and hunt, just as it condemns animals to exist at the pleasure of, and within the boundaries defined for them by, humans.

The predominance of enclosures – the prison where Roberto is held, the boundaries of the archeologists' camps, the corrals in which Francisco raises lowland pacas – subdues the natural world that creeps and descends, seeks the light or flees it, according to biological impulse. The feature film *Lo que soñó Sebastián* / *What Sebastián Dreamt* (2004), directed by Rey Rosa and co-written by him with the American poet Robert Fitterman, which was selected for the Sundance Festival and enjoyed some success on the repertory circuit, depicts these insights in more dramatic strokes. The film is dedicated to the memory of Paul Bowles; the soundtrack includes some of Bowles's musical compositions. Released ten years after the novel's publication, the film confronts a Petén in more acute ecological crisis than that of the novel; concomitantly, the film's techniques are less subtle. The opening credits include information about the region's rate of deforestation. The annual spring burning of the forest to create cattle pastures, which is absent from the novel (because it was far less prevalent in the early 1990s), serves as a repeated image in the film. The voice-over narration provides information about the number of animals killed for each acre of forest that is incinerated. The plot is less complex than that of the book. Roberto is a malevolent figure; when Véronique rejects his advances, he rapes her. The dynamics of power in the jungle have become even more violent and exploitive. The traffic in Mayan artefacts is given greater play in the film. With the end of the civil war, foreign extractive industries, rather than military hunts for guerrillas, are the agents of environmental destruction. Roberto's complaint to Véronique about Sebastián's misplaced priorities is revised. The reference to the army burning "I-don't-know-how-many acres of jungle to scorch a couple of subversives" (Rey Rosa [1994] 200, 76–7) is replaced by an allusion to "the oil company that's ruining the river" (Rey Rosa, dir., 2004). Consistent with globalization's enfeeblement of the national bourgeoisie, the Sebastián of the film is not a Guatemalan, but a wealthy Spaniard who has inherited a house in the Central American jungle. On both sides, the forces struggling over the future of Guatemala's rainforest are no longer national, but global.

From Charting the Landscape to Historical Clarification: Los sordos and La cola del dragón

Lo que soñó Sebastián initiated Rey Rosa's quest to elucidate the injustices of recent Guatemalan history and engage with national society in an era of accelerated globalization. His next novel, The Good Cripplem(1996), represents a coded attempt to engage with the consequences of his mother's kidnapping. Borges's theory that the lives suppressed by an individual's choices, or by turns of fate, remain as present as those that we actually live, underlies this engagement. In The Good Cripple, Rey Rosa presents an alternate version of a family shattered by a kidnapping. The son, rather than the mother, is kidnapped. Rather than paying the ransom, the father refuses to do so. Rather than being returned safely, the hostage is mutilated by the kidnappers. Like his creator, the protagonist moves to Morocco and meets Paul Bowles; in contrast to the author, he marries the girlfriend of his youth and has little to say to the expatriate American writer. In addition, The Good Cripple sketches in themes of bourgeois life in Guatemala City.

This seed germinates in Rey Rosa's next novel, Que me maten si … A work of extreme concision, this novel employs a thriller-like plot to depict how different pieces of Guatemala's geography and history fit together in the aftermath of the civil war (while also including scenes set in England and France). *Que me maten si ...* is a pivotal work in Rey Rosa's career that merits a more extended analysis than is possible here. In defiance of an accelerated globalization that shrivels the nation's claim to define identity, Rey Rosa engages with Bowles's *Up above the World,* in an act of homage that subtly wrests Guatemala's geography from the American's gaze and consolidates it as his own. Like *Up above the World, Que me maten si ...* sets its primary coordinates in the capital and on the Atlantic Coast; it dramatizes the impact of foreigners present in the country on how Guatemalans interact with each other. By naming the country – which Bowles refrains from doing – Rey Rosa obliges himself to take positions regarding its history. The novel's intertextuality, making more evident than is customary that every literary text must be read as "double" (Kristeva 1978, 85), is underscored by the presence of a Bowles-like figure, an adventurous elderly British travel writer named Lucien Leigh (whose name may also allude to the great travel writer Patrick Leigh Fermor [1915–2011]), who "había vivido más de ochenta y cinco años – casi la mitad de los cuales pasó entre extraños y en lugares apartados" (Rey Rosa 1997, 7; who had lived more than eighty-five years, nearly half of them spent among strangers and in remote places). Leigh, whose age coincides with that of Bowles at the time of the novel's composition, escorts the reader through a Guatemala that remains little known to the bourgeoisie of the capital. When Leigh's wife asks Ernesto, the young male protagonist, what he is doing in Chajul, the site of genocidal acts during the civil war, the female protagonist, Emilia, replies, "Quería enseñarle a mi amigo un poco de su propio país" (Rey Rosa 1997, 32; I wanted to show my friend a little of his own country). Leigh's murder two-thirds of the way through the novel leaves Ernesto and Emilia (and, by implication, Rey Rosa) alone to understand their country on their own. Leigh's death may be read less as a parricidal act against a literary father than as a reflection of Rey Rosa's anxiety at Bowles's approaching death (Esposito 2015) and his awareness of his responsibility to develop literary techniques to deepen his exploration of the Guatemalan landscape, and of the rural Guatemala ignored, despised, and exploited by his own social class, while preserving his high-art ideals. In 127 pages, the novel encompasses bourgeois life in the capital, life in the Ixil Triangle in the aftermath of genocide, the lives of foreign expatriates, and the African-descended cultures of the Atlantic Coast. It broaches controversial topics, such as the testimonies of Rigoberta Menchú, the role of Israeli military advisers in supporting the genocide of the late 1970s and early 1980s,[23] and the human trafficking of war orphans. *Que me maten si ...* establishes that an engagement with Guatemala that embraces the entire nation must entail a clarification of the history of the war years. In this novel, the Maya appear as a substantial presence in national life, defined by their own ethnicities and cultures. Their existence as a chain of related communities makes elucidation of the crimes of the war years a moral necessity. Ernesto's conflict with his upper-class mother over the subject of a television report on bodies found in a mass grave raises for the first time the question of impunity and the implementation of justice in the post-war era:

> –Esa gente fue manipulada – dijo la señora. Se levantó para apagar el aparato y volvió a sentarse.
>   –Viste – dijo Ernesto –. Allí había cadáveres de viejos y de niños. Bebés.
>   –Era la guerra. Pero eso pasó hace más de diez años – dijo ella.
>   –Cometieron crímenes de guerra.
>   –Yo no soy quién para juzgar. (Rey Rosa 1997, 15)

> "Those people were manipulated," his mother said. She got up to turn off the television and sat down again.
>   "You see," Ernesto said. "There were bodies of old people and children. Babies."
>   "It was war. But all that happened more than ten years ago," she said.
>   "They committed war crimes."
>   "It's not my place to judge."

Ernesto's mother's attitude foreshadows present-day debates, in which the dominant sectors of the bourgeoisie are united in their creed of genocide-denial. *Que me maten si ...* is the first of Rey Rosa's novels to raise this debate and to pose the bourgeoisie's false consciousness regarding the civil war as the primary obstacle to national reconciliation. The development of these themes in *Los sordos* goes a step further, questioning whether Western judiciary systems are even capable of delivering justice in Guatemala.

Rey Rosa's disenchantment with his own social class is expressed in ways unrelated to the aftermath of the war, in his caustic urban novel *Piedras encantadas*. Though both *Caballeriza* and *Severina* inhabit the same elite sectors of Guatemala City society as the backdrop for gothic tales, in *El material humano* the issues of a society dominated for decades by the secret police dovetails with the question of individual destiny. The discovery, in July 2005, of Guatemala's secret police archives, at a point in history when the classes represented by the character of Ernesto's mother had assumed that the debate over war crimes was dwindling, has assisted in clarifying incidents from the civil war period and re-energized campaigns for redress of injustices from that time (Weld 2014, 29–49). Rey Rosa's artistic response to these events, and to his own attempts to do research in the archives, is unexpected. Eschewing the indirection and restraint of his previous fiction, this novel infringes on the high art tenets of unity and linguistic purity in two directions: through the inclusion of pages of unprocessed police records, some of which are only tangentially related to the central story; and through the use of a diary format, which records the author's daily life in banal, intimate detail: paranoias, fears, dreams, chores, family dinners, trips to literary festivals, conversations with his infant daughter, drinking, ruptures and reconciliations with his partner. The virtual erasure of the boundary between reality and fiction (some individuals are given fictitious names, though in most cases their identities remain obvious) corresponds to the quest of the character "Rey Rosa" as he tries to use the archives to find out who kidnapped his mother in 1981. His conclusion that the answer to such a question "tal vez es sólo para mí" (Rey Rosa 2009, 179; maybe is just for me), while it may appear to weaken the argument for public reconciliation, in fact strengthens it. Rey Rosa's personal need to find out what happened to a member of his family reinforces the right of all citizens to benefit from the historical clarification (this argument is developed by Helen Mack Chang in chapter 5).

El material humano renews Rey Rosa's campaign against the elite's obstruction of the justice system, acting as the forerunner to two works that give a central place to the theme of justice. *Los sordos*, which in 2013 won the Twenty-First Century Award for the best foreign novel published in China, is, at 232 pages, by far the longest of his novels. Exhibiting his customary extreme concision, Rey Rosa enfolds a broad portrait of Guatemalan society into its modest length. The present analysis will focus on the novel's later sections. The protagonist, Cayetano Aguilar, is one of the bodyguards who have become an ubiquitous feature of Central American life. Cayetano is hired to watch over Clara Casares, the idle forty-year-old daughter of a millionaire industrialist. Multiple levels of intertextuality overlap: Rey Rosa plants in the novel literary allusions, themes from his earlier work, and, in a residue of the testimonial techniques of *El material humano*, references, ranging from the discreet to the blatant, to his own life. Themes of imprisonment and human experimentation, broached in *The Pelcari Project* and, in a different vein, in *Caballeriza*, recur;[24] surveillance technology, which is pivotal to *Que me maten si ...* and *El material humano*, returns; the kidnapping themes of *The Good Cripple* and *El material humano* are given an original twist.

The patriarchal figure of the industrialist Claudio Casares is reminiscent of Rey Rosa's father; by giving Don Claudio one of Adolfo Bioy Casares's surnames, Rey Rosa merges his biological and literary father figures. A conversation about lawyers includes a reference to the law firm Salas & Salas, which was founded by Rey Rosa's maternal uncles. A character who appears at the long party scene that sets up the conclusion of the novel's first half, jarring the complacent members of the elite with his iconoclastic opinions and his "vestimenta casual [que] era como una crítica privada al chic de los demás" (Rey Rosa 2012, 75; his casual dress [which] was like a private criticism of the others' chic), is named Sebastián and appears to be the protagonist of *Lo que soñó Sebastián* many years later. Now working for a non-governmental organization that tracks who is getting richer and who is getting poorer, and why, he incarnates attention to the social injustices to which his upper-class contemporaries prefer to remain symbolically "deaf." When Sebastián exits the party, questions are asked about who finances his NGO: "–Los europeos, los gringos, los chinos – dijo el banquero –. Socialistas, es claro" (Rey Rosa 2012, 77; "The Europeans, the gringos, the Chinese," the banker said. "Socialists for sure"). Rey Rosa includes an overt allusion to how such acid commentary on his own social class, combined with his increasingly outspoken espousal of justice for and, more shockingly to his own milieu, *by* the Maya, alienate him from the class in which he was raised. Near the conclusion of *Los sordos*, Clara Casares and her lover Javier discuss whether a writer acquaintance named Rodrigo can be considered "producto de su tiempo y su medio" (a product of his time and his milieu), with Clara expressing the opinion that the writer does not belong to his social class: "Ése está, y siempre estuvo, *fuera* del tiempo – agregó, pero en un tono despectivo, descalificador – y en ningún lugar, si me preguntas a mí" (Rey Rosa 2012, 215; "That guy is, and always was, *outside* time," she added, but in a disdainful, disqualifying tone, "and nowhere at all, if you ask me").

This exchange illustrates Rey Rosa's awareness that while in youth he may have lived outside Guatemala's physical boundaries, in recent years his dissidence from the upper-class notion of civilization, which assumes that Mayan peoples are inferior beings mired in an archaic past, has projected him into a position of no longer sharing commonly held class assumptions. The novel's title poses the physiological deafness of a young Mayan boy, Andrés Curruchich,[25] against the metaphorical deafness of the upper classes. Clara hurls this accusation – "¡Están sordos todos!" (Rey Rosa 2012, 117; You're all deaf!) – at her father in her final phone conversation with him on the night before his death. She and her friends discuss founding a hospital in the countryside as a charitable project. Clara's lover, Javier, is her father's new lawyer. When not working for Don Claudio, he is based in Geneva. During his long absence, their relationship breaks down. His first letters to Clara are passionate; by the end of his absence, he writes, "Entiendo que quieras que dejemos de comunicarnos durante algún tiempo" (54; I understand that you want us to stop communicating for a while). When he returns to Guatemala, he spends the night after the party with Clara, drugs her, and spirits her away to his friend Dr Kubelka's experimental hospital in an architecturally spectacular former "hanging" hotel constructed on stilts overlooking a remote corner of Lake Atitlán. Here he keeps Clara docile, compliant, and willing to remain with him through regular doses of drugs. Every few weeks, Clara makes a brief phone call to her family, pretending to be in a different country each time and casting doubt on whether her disappearance can be classified as a kidnapping. Clara's young bodyguard, Cayetano, who idealizes her as he idealized an unattainable young woman in his home town, vows to locate her. When he finds the hospital, Javier and Dr Kubelka drug and confine him. Cayetano escapes, stealing a laptop, which, in a vein reminiscent of Bioy Casares's *A Plan for Escape* and *Asleep in the Sun*, contains files that reveal that Kubelka has been kidnapping and experimenting on Mayan children: "fotos de niños – las caras demenciales, las cabezas trepanadas – los diagramas de circuitos cerebrales, los nombres de drogas y hormonas (dopamina, oxitocina, paxil, rohypnol), instrumentos (neuromoduladores, microelectrodos) y procedimientos (implantes, avulsiones, trepanaciones)" (168; photos of children – demented faces, trepanned heads – diagrams of circuits of the brain, names of drugs and hormones [dopamine, oxytocin, paroxetine, flunitrazepam], instruments [neuromodulators, micro-electrodes] and procedures [implants, avulsion fractures, trepanations]).

As a lower-class young man from the predominantly mestizo Oriente region of eastern Guatemala, Cayetano is exempt from the reciprocal hostility that pits Europeanized aristocrats like Clara, Javier and

Dr Kubelka against the Maya. He shows the Mayan church minister to whom he takes the laptop the card that licenses him as a police weapons instructor and says, "Soy autoridad" (Rey Rosa 2012, 167; I'm the law). This common Central American expression invokes the weight of institutional authority; to Cayetano's surprise, the Mayan minister replies, "Yo también soy autoridad, hermano" (168; I'm the law too, brother). This exchange crystallizes Rey Rosa's challenge to elite presumption, insisting on Mayan structures of authority as parallel national institutions rather than as folk-superstition. The minister's statement initiates a complex chicken-and-egg game. *Los sordos* opens with an author's note in which Rey Rosa thanks Mayan legal scholars in Nahualá for teaching him about "el milenario sistema de justicia maya" (the thousand-year-old system of Mayan justice), explains Mayan legal terminology, invokes the importance of "jurisprudencia maya" (Mayan jurisprudence), and invokes Nahualá's probable status as the place where *Título de los señores de Totonicapán* (Title of the lords of Totonicapán), which outlines a Mayan concept of social cohesion, was written in the Quiché-Maya language in 1554 (10–11). By emphasizing the authority of such traditions, Rey Rosa both dignifies the Maya and challenges his own class's racism and genocide denial. These projects are inextricably linked. It is possible to interpret the hospital, and Dr Kubelka's experiments on local children, as a microcosm of the genocidal Guatemalan state and its torturing of Mayan communities and to see the Mayan trial to which Clara is subjected as a model of how to render judgment on the Guatemalan state's historical crimes. The seizure of Clara and Javier by the elders of Nahualá inverts, and satirizes, elite assumptions. Having resisted putting generals such as Efraín Ríos Montt on trial in a Western justice system, the elite is now subjected to Mayan justice. Clara, who comes from a family that lives in fear of kidnappings, is accused of kidnapping Mayan children. Javier, the international lawyer, reveals that he is ignorant of the existence of Mayan law and of its recognized legal standing within the territorial limits of Mayan communities. A Mayan judge tells Javier,

En esa tierra coexistían dos formas de derecho. La occidental, o *kaxlán*, y la maya. ¿No lo sabía el licenciado?

Javier dijo que no lo sabía.

El juez continuó: si alguien era detenido como supuesto delincuente dentro de los límites jurisdiccionales de una comunidad determinada, podía optar por ser juzgado por las autoridades mayas, en lugar del Ministero Público. (Rey Rosa 2012, 197)

Two forms of law coexisted in this land. The Western, or *kaxlán*, and the Maya. Didn't the lawyer know that?
Javier said that he didn't know it.
The judge continued: if someone was detained as an alleged offender within the jurisdictional boundaries of a given Mayan community, they could choose to be judged by the Mayan authorities rather than by the Public Prosecutor's Office.

The use of the derisive *kaxlán* – a Quiché and Kaqchiquel term for non-Mayan Guatemalans roughly equivalent to the way some Latin Americans use *gringo* to refer to light-skinned foreigners – reinforces the notion of Western law as an alien construct. The implication that Clara and Javier's arrest serves as symbolic retribution for the elite's defection from the project of an equitable, institutionalized Guatemalan nation, promulgated by the 1996 Peace Accords, is reinforced when the Mayan judge reminds Javier that he himself supported the adoption of the International Labour Organization convention that legalized the use of Mayan law: "Usted mismo, si no recuerdo mal, estaba a favor. A ver. Hice alguna huisachada para el bufete Robles & Rosas. ¿Creo que en el noventa y seis?" (Rey Rosa 2012, 203; You yourself, if I'm not mistaken, were in favour of this. Let's see. I did a bit of informal legal work for the firm of Robles & Rosas. I think in about ninety-six?) The judge's statement, in addition to emphasizing his authority in Western terms, by virtue of his work at Javier's law firm, reprimands the elite for its failure to remember the idealistic stances to which it subscribed in 1996, during the negotiation of the Peace Accords. Though Javier agrees to a Mayan trial, his private thoughts about this process confirm that it is racial prejudice, rather than the egalitarianism proposed by the Peace Accords, that guides his view of his situation. Javier considers the elders of Nahualá as "los portadores del virus de una religión caduca que se resistía a la extinción" (207; the carriers of the virus of a decrepit religion that refused to become extinct). When a Mayan elder holding a staff of office comes to see him in the yard of the hospital, Javier thinks, "Es como un mono" (208; He looks like a monkey). As their interview proceeds, with the elder providing him with the ancient history of the legal system under which he will be judged, Javier becomes increasingly disoriented and uncertain whether his fate has been decided in advance. In fact, the outcome of the trial is ambiguous. Clara's explanation that Andrés Curruchich, the deaf boy who disappears in the novel's opening scene, has been given the capacity to hear by experimental surgery performed in the hospital, prompts the judge, Juan Chox, to recall early Mayan practices of performing trepanations to cure mental illness. Exemplifying a cross-cultural understanding that the Ladinos do not match, Chox notes, "Dejamos de practicarlas hace tiempo … Esas cosas no están bien. Sin embargo nos causa gran alegría comprobar que Andrés haya sido curado de su sordera en el hospital de ustedes" (220; We stopped practising them [trepanations] a long time ago … These things are not right. Nevertheless it causes us great joy to confirm that Andrés has been cured of his deafness in your hospital). Yet, it is noted, in Mayan culture the deaf have an exalted role and special authority; the Maya even have their own form of sign language. By "curing" Andrés, the *kaxláns* have deprived him of this status.

Cayetano, the novel's principal narrative consciousness, is bidden to leave the court as Chox is about to pronounce the Mayan elders' verdict; Clara's fate, like that of Javier, remains unknown. The narrator follows Cayetano away from Lake Atitlán and back to the Oriente region, where he makes a fatal attempt to win back the local girl he longed for as a youth from her drug trafficker husband. The failure of Mayan justice, and Mayan culture in general, to be acknowledged by national institutions is underlined when, over the next few days, Cayetano listens to the regional news to find out what happened to Clara, and "no aparecía nada sobre el hospital colgante junto al lago ni sobre las autoridades mayas de Nahualá" (Rey Rosa 2012, 223; no news appeared about the hanging hospital next to the lake or the Mayan authorities of Nahualá).

The fact that the outcome of Clara and Javier's case remains unknowable makes clear that Mayan justice remains pending business in Guatemala. These characters from elite backgrounds consent to a Mayan trial only because their lives are threatened by an enraged mob. The issue of how to bring the genocide-denying elite to acknowledge the facts of Guatemalan history and the presence of a non-Western cultural tradition that exerts a primary claim to the land becomes the central theme of Rey Rosa's collection of polemical journalism, *La cola del dragón: No ficciones*. The subtitle, both a Hispanicization of the English term *non-fiction* and an implicit diametrical opposite, in both genre and ideology, to *Ficciones* ([1944] 1971), the Borges collection that inspired Rey Rosa to become a writer, underlines this book's rupture with his aesthetic commitments. One of the anomalies that, until recently, set Rey Rosa apart from the Spanish American tradition was that he eschewed the two-hundred-year-old role of the Latin American writer as an authority on national destiny who disseminates his views through newspaper columns, essays, speeches, or active engagement in politics (Castañeda 1993, 177–8). For a quarter of a century, Rey Rosa spoke, elliptically and enigmatically, solely through his fiction. Two-thirds of the essays in *La cola del dragón* were written in 2013 and 2014; excluding pieces on Paul Bowles and Miquel Barceló, only one essay is from before 2011. This flurry of journalistic activity, coinciding with the trial of Efraín Ríos Montt, illustrates Rey Rosa's growing frustration with the refusal of Guatemalan opinion-makers not only to address social injustice, but even to concede that it exists. The book's title comes from words spoken to Rey Rosa by a peasant in the Ixil Triangle, the epicentre of the genocide in the highlands during the civil war, who attributes the long delay in bringing Ríos Montt to trial to the people's fear of stirring up the past: "No hay que tocarle la cola al dragón" (Rey Rosa 2014a, 74; You mustn't touch the dragon's tail).

Touching, or even treading on, the dragon's tail is the central activity of this collection, both in the sense of confronting the past and of breaking down the ideological fantasy-land inhabited by the Guatemalan elite. Readers of Rey Rosa's fiction, who are accustomed to his narrative restraint, may be surprised by the political commitment and outspoken disenchantment with the structures of post-1996 society, both in Guatemala and in the international realm, expressed here. The title essay, in good Bowlesian style, tries to understand the Ríos Montt trial through a journey into the Ixil region. It is a mixture of feature-style reportage, nourished by interviews with witnesses whose names have been changed, with bracing travel writing. The details of the systematic extermination of one-third of the Ixil-Maya ethnicity in the early 1980s, and of the intimidation that continues to reduce local elections to events with pre-programmed outcomes, alternate with passages of lyrical descriptive writing whose tone and imagery draw on Rey Rosa's years in Europe: "el camino entre Nebaj y Cotzal dejaba ver toda la belleza de un valle por encima de las nubes. El bizarro perfil de las montañas, acentuado por cortinas de niebla finísima, había sido delineado por un artista de inspiración gaudiesca; la carretera de asfalto negro se ceñía al terreno curvilíneo como un listón" (Rey Rosa 2014a, 76; the road between Nebaj and Cotzal revealed the valley above the clouds in all its beauty. The generous profile of the mountains, accentuated by curtains of very fine mist, had been sketched by an artist inspired by Gaudí; the black asphalt highway hugged the curved terrain like a ribbon). The allusion to the Barcelona modernist architect Antoni Gaudí (1852–1926), echoing the allusion to the *pietà* in *Lo que soñó Sebastián*, has the contradictory effect of reinforcing Rey Rosa's status as an observer rather than a participant in the history he investigates; the allusions, which separate him from the testimonio genre of colloquial, politically committed confession, heighten his credibility by maintaining his distance. In a similar vein, he refutes the elite's lies by citing declassified documents from the unimpeachable (from an elite point of view) source of the U.S. State Department (67–8). A literary language that, in the Borgesian tradition, approaches the local through a universal culture, out-manoeuvres the ruling classes, whose genocide-denial is based on their parochial assumption of superiority to the *indios*. Rey Rosa is careful to delineate the full dimensions of the genocide: not merely mass killings and razing of towns and villages, but also the policies that were designed to ensure that surviving children were unable to sustain the Ixil-Maya language and culture: "los niños ixiles sobrevivientes fueron traslados de un grupo étnico a otros distintos, y reubicados forzosamente, a veces para formar parte de familias de los mismos soldados que exterminaron a sus familias ... de modo que su pasado quedó erradicado" (81; the surviving Ixil children were transported from one ethnic group to others that were different, and forcibly relocated, at times to become part of the families of the very soldiers who exterminated their families ... so that their past was eradicated). The restraint of Rey Rosa's fiction is reflected in his strictness with evidence. His informants are equally circumspect. He quotes the Ixil elders' terse statement when Ríos Montt was brought to trial: "Aprobamos que se haya hecho justicia. Es el primer paso" (82; We approve of the fact that justice has been done. It is the first step).

The step beyond bringing an octogenarian general to trial (even though Guatemala's Constitutional Court later quashed his conviction) lies in achieving nationwide recognition of the history of the civil war. This second step is the necessary precondition for social justice, since only once the genocide is acknowledged will the worth of Mayan peoples as fellow citizens become axiomatic; only then can policies be formulated to promote a just order in light of this equality of value. The incessant repetition of the slogan "No hubo genocidio" (There was no genocide) by every post-1996 president except Álvaro Colom (2008–12), and by dozens of Guatemalan government and military officials, diplomats, newspaper journalists, and television commentators, is the mindless morass into which Rey Rosa must descend in order to corroborate the history he wishes to make available by touching the dragon's tail. Much of *La cola del dragón* consists of the uncomfortable spectacle of a lucid intellectual going to great pains to refute the propagandistic con-coctions of mediocrities. One of the Guatemalan elite's most farcical responses to the genocide is the counter-assertion that the only geno-cide in Guatemalan history occurred when the liberal government of Jacobo Arbenz (1951–4), shortly prior to being overthrown by a U.S.-orchestrated invasion force (Schlesinger and Kinzer 1982, 191–226), as analysed by Candace Johnson in chapter 1, executed a right-wing journalist and a number of his followers in Antigua. This incident, brandished by the Guatemalan right whenever the topic of genocide is raised, is debunked in meticulous detail in a chapter co-written by Rey Rosa and the journalist Sebastián Escalón. The fact that it is even necessary to refute the contention that a single summary execution is far worse than eliminating one-third of an ethnic group's population (among many other genocidal acts) exposes the poisoned intellectual environment of contemporary Guatemala. In a similar vein, Rey Rosa must refute standard upper-class red herrings, such as the assertion that the Maya existed in the remote past, as the Roman Empire existed, but that in the contemporary context "Maya" is simply a cover identity adopted by degenerate criminals who wish to "vivir comodamente de ingenuos guatemaltecos y sus 'conflictos'" (Rey Rosa 2014a, 139; live easily off innocent Guatemalans and their "conflicts"). The "criminal" reference is an upper-class catch-all that refers to Mayan efforts to reclaim land stolen from them by military officers and the elite dur-ing the civil war, opposition to the annexation of their land by foreign extractive industries, and the fact that many members of the criminal gangs known as *maras* are of Indigenous extraction – even though the young men who join gangs are overwhelmingly those who, having lost touch with Mayan culture, are seeking alternate forms of belonging. In one of the book's central essays, "La violencia que generamos" (The violence we generate), Rey Rosa quotes a recent German ambassador to Guatemala who, in a speech, suggested that Germany and Guate-mala were both countries that must confront a history of genocide. The ambassador recommended that Guatemalans engage in their own version of the Nuremberg trials: "Hoy en día en Alemania – nos hizo recordar – negar o justificar esos crímenes es a su vez un delito, un acto perseguido penalmente" (Rey Rosa 2014a, 149; Today in Germany, he reminded us, denying or justifying these crimes is itself an offence, an act pursued by the justice system).

Guatemala cannot mount its own Nuremberg trials because, to adopt the terms of the ambassador's analogy, in Guatemala City the Nazis remain in power. Even the rare convictions that are extracted from the justice system, like that of Ríos Montt, risk being overturned. The analogy, and above all the contrast, between Germany and Guatemala inculpates the ruling castes and criminalizes their activities, in the present, as well as the past. The sole longer piece in this book that dates from before 2011, "El tesoro de la sierra" (The treasure of the mountains), published in 2004, is an account of a visit to Honduras to witness the damage caused by the first Canadian mining companies to enter the isthmus, then to the first Canadian mines in Guatemala; a postscript added in 2014 cites a report by the Working Group on Mining and Human Rights in Latin America (2014), which lists the human rights violations, including violent death, visited upon Guatemalans or Hondurans who oppose the implantation of Canadian open-pit mines on their land (Rey Rosa 2014a, 185–6). (A chronology of the incursion of Canadian mining into Guatemala appears in chapter 4 by Magalí Rey Rosa.) Mining becomes the extension of the genocide of the 1980s, not only because it continues the pulverization of Mayan culture, in this case by literally demolishing mountains and valleys on which the Maya depend for their livelihood, cultural identity, and linguistic cohesion, but also because it projects the murderous elite ideologies of the 1980s into the present. The implication is that economic and social neoliberalism, by virtue of being the creed of the elite, is a genocidal belief system. This connection might not be automatic in other Latin American countries; yet Guatemala's particular history converts neoliberal policies into the twenty-first-century extension of the belief systems that underlay the genocide of the 1970s and 1980s. This is evident in Rey Rosa's distinction between passive complicity with violence, which may be found in many citizens, and "la complicidad activa de un empresario de la vieja guardia, la de un militar de alto rango, la de un funcionario medio del sistema financiero o la de un profesor de economía de universidad privada" (139; the active complicity of an old-style businessman, a high-ranking military officer, a middle manager in the financial system, or an economics professor at a private university). Mario Sandoval Alarcón, vice-president of Guatemala, 1974–8, stated, "Si debo deshacerme de la mitad de Guatemala para que la otra mitad pueda vivir en paz, lo haré" (140; If I have to get rid of half of Guatemala so that the other half can live in peace, I'll do it). Those who, like the economics professor at the private university, perpetuate the ideology of free markets, which is the successor to the racist exterminations of the civil war, are also culpable; they, and their class, are drenched in an ideology whose central achievement is genocide, actively implemented in the recent past and continued by alternate means, such as "free market"–justified mining, in the new millennium.[26] The goal remains the same: the extinction of cultures that offend them; this "cultural" imperative takes precedence over profit, since the negligble royalties paid by mining companies to the government ensure that a tiny number of Guatemalans actually benefit from mining. Mining satisfies the elite because it provides them with the opportunity to interact with rich white gringos in what they perceive as an equal partnership, assuaging their cultural and racial insecurities; by breaking up communities and smashing mountains, it promises to complete the unfinished obliteration of Mayan culture that was begun in the 1980s.

Failure to recognize the fact of genocide is key to the stagnation of both human and environmental justice in Guatemala. Having refuted some of the elite's more ridiculous assertions, Rey Rosa closes *La cola del dragón* with thirty pages of excerpts from the CEH report. In the interests of balance, he begins by quoting the report's account of a massacre committed in 1987 by Marxist ORPA guerrillas – even though the two major reports on human rights violations during the war years estimate that 3–5 per cent of the killings of civilians were committed by the guerrillas, and over 80 per cent by the military (Goldman 2007, 22). The final pages document military massacres; the concluding lines, consisting of citations from the CEH report's conclusions, sum up the military's actions with the words "lo que configura un acto de carácter genocida" (Rey Rosa 2014, 249; which constitutes an act of a genocidal nature). By making "genocidal" the book's final word, Rey Rosa establishes this history as the core to which all other themes return. An example occurs in "La caja de los truenos" (The thunder box), where the case of the lawyer Rodrigo Rosenberg, who arranged his own death in order to cast blame on former President Colom (see chapter 7 by Rita M. Palacios), becomes another object lesson on the impossibility of maintaining functioning judicial institutions in the face of elite non-cooperation.

The decision to place "El tesoro de la sierra," written almost ten years earlier, after the more recent articles inspired by the Ríos Montt trial underlines the fact that the environmental destruction that dominates the present is the consequence of the actions of the same elite that covers up the genocide of the past; the mining essay concludes with Rey Rosa questioning two mining executives of his own social class over lunch. This is the culmination of his return to Guatemala: his journeys through the Petén and the Ixil turn him back upon the elite, among whom he was raised, with a highly critical eye. As *Los sordos* and *La cola del dragón* illustrate, the intersection of the civil war and mining is also the intersection of Rey Rosa's modernist restraint and his need to engage. Pairings of both theme and approach, like the pairing of fiction and journalism, cohere around the unresolved dilemma of the Mayan peoples, the truth of their historical experience, and the form their citizenship will take in a future Guatemala.

NOTES

1 Book titles of works written in Spanish are in the original, with English translations in parentheses. An italicized English title refers to an existing translation; a non-italicized English title is a literal rendering of the Spanish title of a book for which no English translation existed at the time of writing. Citations from works written in Spanish are in the original. Where a published translation exists, the English quote is from the translation; otherwise the translation is by the author. The author is grateful to Rodrigo Rey Rosa for correcting errors of chronology and nomenclature in an earlier draft of this chapter.

2 *Cárcel de árboles* (1991; *The Pelcari project*), originally published as a freestanding novel, is included as a short story in *1986: Cuentos completos* (2014; Complete short stories); *El tren a Travancore* (2002; The train to Travancore), published as a travel book, was repackaged as fiction by its inclusion in *Tres novelas exóticas* (2016; Three exotic novels); a conversation with a bullfighter entitled "Entrevista en Ronda" (Interview in Ronda) appears as journalism in *La cola del dragón* and as a short story in *1986*; "Desventajas de la santidad" ("Disadvantages of sainthood") is presented as the transcription of an interview by the author, but is actually fiction.

3 In Central America, testimonio literature was most closely identified with works such as *Me llamo Rigoberta Menchú y así me nació la conciencia* ([1982] 2007; *I, Rigoberta Menchú: An Indian Woman in Guatemala*, 1985), narrated by the Mayan activist Menchú and edited by Elisabeth Burgos-Debray; and the Nicaraguan guerrilla commander Omar Cabezas's *La montaña es algo más que una inmensa estepa verde* (1982; *Fire from the Mountain: The Making of a Sandinista*, 1985). Catherine Nolin, in chapter 2, and Rita M. Palacios in chapter 7, employ the term *testimonio* in slightly different ways.

4 See the Salvadoran Manlío Argueta's *Un día en la vida* ([1980] 1987; *One Day of Life*, 1983) or, in a more complex, postmodern vein, Sergio Ramírez's *Sombras nada más* (2002; Only shadows remain) (Henighan 2014, 517–38).

5 For example, Monterroso (2011), Boldy (2012), Albizúrez Gil (2013), Gutiérrez Mouat (2013).

6 The Guatemalan Marco Antonio Flores's novel *Los compañeros* (1976)
   and the Salvadoran Horacio Castellanos Moya's *La diáspora* ([1989] 2002)
   dramatize the intrigues of Central American guerrillas in Mexico City.
   The generation of Guatemalan intellectuals who were exiled after the 1954
   military coup, such as the sociologist, novelist, and former cabinet minister
   Mario Monteforte Toledo (1911–2003), and the short story writer Augusto
   Monterroso (1922–2003), had also chosen exile in the Mexican capital. Rey
   Rosa did not belong to any political organization. The autobiographical
   protagonist of *El material humano* relates that "aunque nunca tuve vínculos
   directos con ninguna de las organizaciones revolucionarias, mis simpatías
   estaban con ellas y no con el Gobierno" (Rey Rosa 2009, 91–2; although I
   never had direct connections with any of the revolutionary organizations,
   my sympathies were with them and not with the government).

7 Roberto Bolaño (1953–2003), the most influential writer of the post-boom
   generations, praised the Guatemalan and helped to spread his mystique
   by publishing an article about a gruelling journey Rey Rosa made into the
   interior of Mali (Bolaño 2004, 199–200).

8 Though this is not the topic of the present study, Rey Rosa's autobiographical novel *El material humano* makes clear that his mother's kidnapping on
   28 June 1981, probably by a minor guerrilla organization of little strategic significance (she was released, after the family paid a ransom, on 23
   December of the same year) (Rey Rosa 2009, 89), accentuated his alienation
   from Guatemalan reality and his conflicted attitude towards identification
   with his country, encouraging him to remain abroad.

9 *The Sheltering Sky* sold 40,000 copies in hardcover in 1949 and an additional
   200,000 copies in paperback in 1951 (Sawyer-Lauçanno 1989, 287, 305).
   Bowles's less commercially successful later books were published by avant-
   garde presses. As Gore Vidal recalls, "He was entirely out of print in 1979,
   when I wrote an introduction to his collected short stories" (1995, 214).

10 Borges became a point of connection once Rey Rosa discovered that Paul
   Bowles had published the first English translation of a Borges short story. "The
   Circular Ruins," Bowles's rendering of "Las ruinas circulares," appeared in *View*
   magazine in New York in January 1946 (Halpern 2002, 913; Rey Rosa 2014a, 22).

11 When he left for New York, Rey Rosa had given a short story, "El monas-
   terio" (The monastery) to a friend, who submitted it to the Guatemala City
   daily newspaper *El Imparcial*, where it was published (Gray 2007, 164).

12 Since class discussion was in English, Rey Rosa left Tangier to travel: "I left
   some of my stories with him, and I went. When I came back, I went to see

him at his apartment, to say goodbye I guess, and he asked me if he could translate my stories and give them to a publisher" (Gray 2007, 165).

13 The short story collections *The Beggar's Knife*, published by City Lights Books in San Francisco in 1985, and *Dust on Her Tongue*, published by Peter Owen in London in 1989 and by City Lights in 1992; and the short novel *The Pelcari Project*, published in London by Peter Owen in 1991 and in California by Cadmus Editions in 1997.

14 The correct name of this neighbourhood is El Charf.

15 Paul and Jane Bowles took their honeymoon in Guatemala in 1938 and lived in Mexico in 1940 and 1941; Paul Bowles returned to Central America with friends in 1945. He was well-read in Indigenous Mesoamerican mythology (Halpern 2002, 911–13), and both he and his wife wrote fiction set in Guatemala, including Jane Bowles's story "A Guatemalan Idyll" (Bowles 1989, 25–62) and her husband's story "Call at Corazón" (Bowles 2001, 36–47), among others, as well as his novel *Up above the World* ([1966] 2009).

16 "Gorevent" (2014), a story Rey Rosa wrote long after Bowles's death, invokes the Beat writer William S. Burroughs (1914–97), as an example of "escritores que vivieron al margen de la ley" (Rey Rosa [1986] 2014, 436; "writers who lived on the fringes of the law").

17 When he was forced to leave Paris for economic reasons in 1933, Asturias returned home to spend eleven years in drunken, unproductive misery in the same elite Guatemala City circles in which he had been raised. His later life, during which he recovered his creativity and published his major novels, *El Señor Presidente* ([1946] 1948; *The President*, 1967b), *Hombres de maíz* ([1949] 1972; *Men of Maize*) and *Mulata de tal* (1963; *Mulatta and Mr Fly*, 1967a), was spent abroad, mainly in Paris and Buenos Aires, perpetuating the Ladino nationalist image of the Maya that he had consolidated in the 1920s.

18 *El cuchillo del mendigo*, the Spanish original of *The Beggar's Knife*, and *El agua quieta* (Still waters), which Bowles translated as *Dust on Her Tongue*, were published by tiny Publicaciones Vista in Guatemala City in 1986 and 1990 respectively, in each case one year after English publication; *Cárcel de árboles* (*The Pelcari Project*) was published by Fundación Guatemalteca para las Letras in 1991, coinciding with the English version's publication in London (Flores).

19 In *El material humano*, the narrator is reading *Borges* (2006), Bioy Casares's 1600-page memoir of his fifty-year friendship with the master (Rey Rosa 2009, 55). Rey Rosa describes this book as "one of my richest and most amusing reading experiences of the last few years" (Esposito 2015).

20 Rey Rosa credits Bowles with helping him overcome his early Borges-inspired snobbery that longer fiction was not worth reading: "Under Paul Bowles's influence, I began reading a lot of novels, mostly from the

Anglo-Saxons. Conrad, James, Greene, Orwell, Compton-Burnett, Chandler, Highsmith" (Esposito 2015). This selection of English-language literature reflects Bowles's generational and expatriate tastes.

21 For the sexual dynamics of the Borges–Bioy Casares friendship, see Balderston 2012, 67–70.

22 The margay is a small wild cat related to the ocelot, exceedingly rare and on the verge of extinction.

23 After 1977, when U.S. President Jimmy Carter cut off lethal aid to Guatemala in retaliation for human rights abuses, Israel became Guatemala's largest source of military equipment and advisers. Benedicto Lucas García, one of the army's principal military commanders during the civil war, reports that he informed his brother, President Romeo Lucas García (1978–82), "Nosotros tenemos la ayuda de Israel. No necesitamos la ayuda de Estados Unidos" (Gutiérrez Valdizán and Serrano Echeverría 2016; "We have the aid of Israel. We don't need the aid of the United States"). The Israeli economic presence, which during the Civil War years included a Galil munitions factory in Alta Verapaz, persisted in post-1996 Guatemala, being particularly salient in the area of private security firms. On the war years, see Hunter (1987, 111–35); Dunkerley (1988, 489–90); Cockburn (1991, 216–22).

24 Rey Rosa would return to theme of imprisonment in his short story, "1986," written in 2013, which, as the use of the date as a title indicates, reprises the ambiance of *The Pelcari Project*, this time in an explicitly identified Guatemalan location and historical moment (Rey Rosa 2014b, 399–427).

25 As Stephen Henighan and Candace Johnson explain in Chapter Ten, this character's name coincides with that of one of Guatemala's most famous painters. This coincidence does not appear to be intentional.

26 Expressions of a genocidal project on the part of the Guatemalan elite can be disarmingly unvarnished. In 1995, one U.S.-educated member of the elite told the author of this chapter, in flawless English: "I have a beautiful country. I deserve a better people. We need to replace our people."

WORKS CONSULTED

Alazraki, Jaime. 1990. "¿Qué es lo neofantástico?" *Mester* 19 (2): 21–33.

Albacete, Juanjo. 2014. "Pisar la cola del dragón: Conversación con Rodrigo Rey Rosa." De Verdad TV, 28 October. http://deverdaddigital.com/articulo/19878/pisar-la-cola-al-dragon.

Albizúrez Gil, Mónica. 2013. "«El material humano» de Rodrigo Rey Rosa: El archivo como disputa." *Centroamericana* 23 (2): 5–30.

Argueta, Manlío. (1980) 1987. *Un día en la vida*. San José: EDUCA.
– 1983. *One Day of Life*. Translated by Bill Brow. New York: Random House.
Arias, Arturo. 2007. *Taking Their Word: Literature and the Signs of Central America*. Minneapolis: University of Minnesota Press.
Asturias, Miguel Ángel. (1946) 1948. *El Señor Presidente*. Buenos Aires: Editorial Losada.
– (1949) 1972. *Hombres de maíz*. Madrid: Alianza Editorial.
– 1963. *Mulata de tal*. Buenos Aires: Editorial Losada.
– 1967a. *Mulatta and Mr Fly*. Translated by Gregory Rabassa. London: Peter Owen.
– 1967b. *The President*. Translated by Frances Partridge. London: Victor Gollancz
– 1975. *Men of Maize*. Translated by Gerald Martin. New York: Delacorte.
Balderston, Daniel. 2012. "Borges's Appendix: Reflections on Bioy's Diary." In *Adolfo Bioy Casares: Borges, Fiction, Art*, edited by Karl Posso, 59–72. Cardiff: University of Wales Press.
Bertolucci, Bernardo, dir. 1990. *The Sheltering Sky*. With Debra Winger, John Malkovich, Campbell Scott, Jill Bennett. Jeremy Thomas Productions. Warner Brothers.
Beverley, John. 1993. *Against Literature*. Minneapolis: University of Minnesota Press.
Bioy Casares, Adolfo. (1940) 1982. *La invención de Morel*. *La invención de Morel: El gran Serafin*. Edited by Trinidad Barrera. Madrid: Cátedra.
– (1945) 1969. *Plan de evasión*. Buenos Aires: Galerna.
– 1973. *Dormir al sol*. Buenos Aires: Emecé.
– 2004. *Asleep in the Sun*. Translated by Suzanne Jill Levine. New York: New York Review Books.
– 2006. *Borges*. Edited by Daniel Martino. Barcelona: Destino.
Bloom, Harold. 1973. *The Anxiety of Influence: A Theory of Poetry*. New York: Oxford University Press.
Boland Osegueda, Roy C. 2005. "The Central American Novel." In *The Cambridge Companion to the Latin American Novel*, edited by Efraín Kristal, 162–80. Cambridge: Cambridge University Press. https://doi.org/10.1017/CCOL0521825334.009.
Bolaño, Roberto. 2004. *Entre paréntesis: Ensayos, articulos y discursos (1998–2003)*. Edited by Ignacio Echevarría. Barcelona: Editorial Anagrama.
Boldy, Steven. 2012. "Political Violence Revisited: Intellectual and Family in Five Latin American Novels, 2006 to 2009." *Forum for Modern Language Studies* 48 (3): 336–50. https://doi.org/10.1093/fmls/cqs015.
Borges, Jorge Luis. (1944) 1971. *Ficciones*. Madrid: Alianza Emecé.
Bowles, Jane. 1989. *Everything Is Nice: The Collected Stories*. London: Virago.

Bowles, Paul. (1966) 2009. *Up above the World*. London: Penguin Modern Classics.

– (1991) 1997. "Afterword." In *Rey Rosa, Rodrigo, The Pelcari Project*, translated by Paul Bowles, 118–19. Tiburon, CA: Cadmus Editions.

– 1994. *In Touch: The Letters of Paul Bowles*. Edited by Jeffrey Miller. New York: Farrar, Straus and Giroux.

– 2001. *Collected Stories*. Introduction by James Lasdun. London: Penguin Modern Classics.

– 2002. *The Sheltering Sky; Let It Come Down; The Spider's House*. Edited by Daniel Halpern. New York: Library of America.

Burgos-Debray, Elisabeth, ed. (1982) 2007. *Me llamo Rigoberta Menchú y así me nació la conciencia*. México: Siglo Veintiuno.

– 1985. *I, Rigoberta Menchú: An Indian Woman in Guatemala*. Translated by Ann Wright. London: Verso.

Cabezas, Omar. 1982. *La montaña es algo más que una inmensa estepa verde*. Managua: Editorial Nueva Nicaragua.

– 1985. *Fire from the Mountain: The Making of a Sandinista*. Translated by Kathleen Weaver. Foreword by Carlos Fuentes. Afterword by Walter LaFeber. New York: Crown Publishers.

Caponi, Gena Dagel, ed. 1993. *Conversations with Paul Bowles*. Jackson: University of Mississippi Press.

Carr, Virginia Spencer. 2004. *Paul Bowles: A Life*. New York: Scribner.

Castañeda, Jorge G. 1993. *Utopia Unarmed: The Latin American Left after the Cold War*. New York: Random House.

Moya, Horacio Castellanos. (1989) 2002. *La diáspora*. San Salvador: UCA Editores.

– 1997. *El asco: Thomas Bernhard en San Salvador*. San Salvador: Editorial Arcoiris.

– 2016. *Revulsion: Thomas Benhard in San Salvador*. Translated by Lee Klein. New York: New Directions.

Cockburn, Andrew, and Leslie Cockburn. 1991. *Dangerous Liaison: The Inside Story of the U.S.–Israel Covert Relationship*. New York: Harper Collins.

Dunkerley, James. 1988. *Power in the Isthmus: A Political History of Modern Central America*. London: Verso.

Esposito, Scott. 2015. "Interview with Rodrigo Rey Rosa." *White Review*, January.http:/// www.thewhitereview.org/feature/interview-with-rodrigo-rey-rosa/.

Faulkner, William. (1936) 1964. *Absalom, Absalom!* New York: Modern Library.

– (1939) 1962. *The Wild Palms*. London: Chatto & Windus.

Flores, Marco Antonio. (1976) 2006. *Los compañeros*. Guatemala: F&G Editores.

Flores, Ronald. n.d. "The Enigmatic Drifter." *Latin American Review of Books*. www.latamrob.com/the-enigmatic-drifter/.

García Márquez, Gabriel. 1967. *Cien años de soledad*. Buenos Aires: Editorial Sudamericana.

– 1970. *One Hundred Years of Solitude*. Translated by Gregory Rabassa. New York: Harper & Row.

Gellhorn, Martha. 1989. *The View from the Ground*. London: Granta Books.

Goldman, Francisco. 2007. *The Art of Political Murder: Who Killed the Bishop?* New York: Grove.

– 2013. "Rodrigo Rey Rosa." Translated by Ellie Robins. *Bomb* 125 (Fall). www.bombmagazine.org/articles/rodrigo-rey-rosa/.

Gray, Jeffrey. 2007. "Placing the Placeless: A Conversation with Rodrigo Rey Rosa." *A Contracorriente: una revista de historia social y literatura de América Latina* 4 (2): 160–86.

Gutiérrez Mouat, Ricardo, 2013. "El lenguaje de los derechos humanos en tres obras de ficción: *La muerte y la doncella, Insensatez y El material humano*." *A Contracorriente: Una revista de historia social y literatura de América Latina* 11 (1): 39–62.

Gutiérrez Valdizán, Alejandra, and Julio Serrano Echeverría, dirs. 2016. *Benedicto*. Vimeo. https://vimeo.com/179519938.

Halpern, Daniel. 2002. "Chronology." In Bowles, *The Sheltering Sky; Let It Come Down; The Spider's House*, edited by Daniel Halperin, 905–24. New York: Library of America.

Henighan, Stephen. 1999. *Assuming the Light: The Parisian Literary Apprenticeship of Miguel Ángel Asturias*. Oxford: Legenda.

– 2012. "Every Man Is an Island: Bioy's Fiction." In *Adolfo Bioy Casares: Borges, Fiction, Art*, edited by Karl Posso, 89–112. Cardiff: University of Wales Press.

– 2014. *Sandino's Nation: Ernesto Cardenal and Sergio Ramírez Writing Nicaragua, 1940–2012*. Montreal and Kingston: McGill-Queen's University Press.

Hunter, Jane. 1987. *Israeli Foreign Policy: South Africa and Central America*. Boston: South End.

Kristal, Efraín, ed. 2005. *The Cambridge Companion to the Latin American Novel*. Cambridge: Cambridge University Press. https://doi.org/10.1017/CCOL0521825334.

Kristeva, Julia. 1978. *Sémeiotiké: Recherches pour une sémanalyse*. Paris: Éditions du Seuil.

LaFeber, Walter. 1983. *Inevitable Revolutions: The United States in Central America*. New York: W.W. Norton.

Martin, Gerald. 2008. *Gabriel García Márquez*. London: Penguin.

Monterroso, Augusto. 2011. "Yo, el protagonista: La autoficción en una novela de Rodrigo Rey Rosa." *Centroamericana* 20:119–28.

Rama, Ángel. 1982. *La novela en América Latina: Panorama 1920–1980*. Bogotá: Procultura/Instituto Colombiano de Cultura.

Ramírez, Sergio. 2002. *Sombras nada más*. D.F., México: Alfaguara.

Rey Rosa, Rodrigo. 1985. *The Beggar's Knife*. Translated by Paul Bowles. San Francisco: City Lights Books.

- 2014. *1986*. Barcelona: Alfaguara.

- (1989) 1992. *Dust on Her Tongue*. Translated by Paul Bowles. San Francisco: City Lights Books, 1992.

- 1991. *Cárcel de árboles/El Salvador de Buques*. Barcelona: Seix Barral.

- (1991) 1997. *The Pelcari Project*. Translated by Paul Bowles. Tiburon, CA: Cadmus Editions.

- (1994) 2000. *Lo que soñó Sebastián*. Guatemala: Magna Terra Editores.

- 1996. *El cojo bueno*. Madrid: Santillana, SA.

- 1997. *Que me maten si …* Barcelona: Seix Barral.

- 1999. *La orilla africana*. Barcelona: Seix Barral.

- 2001. *Piedras encantadas*. Barcelona: Seix Barral.

- , dir. 2004. *Lo que soñó Sebastián / What Sebastian Dreamt*. With Andoni Gracia, Juliette Deschamps, Juan Carlos Vellido. Productora El Escarbudo. Vimeo. https://vimeo.com/19284551.

- 2006. *Caballeriza*. Barcelona: Seix Barral.

- 2009. *El material humano*. Barcelona: Editorial Anagrama.

- 2011. *Severina*. D.F., México: Alfaguara.

- 2012. *Los sordos*. D.F., México: Alfaguara.

- 2014a. *La cola del dragon: No ficciones*. Valencia: Ediciones Contrabando.

- 2014b. *Severina*. Translated by Chris Andrews. New Haven, CT: Yale University Press.

- 2016. *Tres novelas exóticas*. Barcelona: Alfaguara.

Sawyer-Lauçanno, Christopher. 1989. *An Invisible Spectator: A Biography of Paul Bowles*. New York: Weidenfeld and Nicholson.

Schlesinger, Stephen, and Stephen Kinzer. 1982. *Bitter Fruit: The Untold Story of the American Coup in Guatemala*. Garden City, NY: Doubleday.

Vargas Llosa, Mario. 1965. *La casa verde*. Barcelona: Seix Barral.

Vidal, Gore. 1995. *Palimpsest: A Memoir*. New York: Random House.

Weld, Kirsten. 2014. *Paper Cadavers: The Archives of Dictatorship in Guatemala*. Durham, NC: Duke University Press. https://doi.org/10.1215/9780822376583.

Williamson, Edwin. 2004. *Borges: A Life*. New York: Viking Penguin.

www.dplf.org/sites/default/files/report_canadian_mining_executive_summary.pdf.

9 Press Clippings: The Daily News in Guatemala

W. GEORGE LOVELL

The sleepy café I frequented during the war years no longer exists. It was its very own combat zone one rainy morning when the girl-friend of the owner, a Guatemalan of Italian descent, found out that he had been cheating on her. Drenched to the skin, she burst through the door with a basket full of stones, which she began to hurl at him. My cappuccino was put on hold while the startled miscreant spun around, dodging the missiles that crashed into the mirrored gantry behind him. The assault lasted a calamitous minute, my command of Spanish blasphemy increasing with each throw. After the mayhem was over, the owner looked at me, raised his eyebrows, and shrugged. He then served me coffee rather than attend to the fallout of spilled liquor and broken glass. He nodded at the table where I had the newspapers spread out.

"As bad as yesterday?" he inquired.
"Even worse," I replied.
"I'll leave you to it."

So he left me to it, reading the news of Guatemala as it was reported in Guatemala that particular day – and for countless others during my visits to the country between 1974 and 2015. Vestiges of a war that began in 1960 and that formally ended with Peace Accords in 1996 not only linger but have spawned their own post-conflict predicaments. These are often as challenging for journalists to engage as the ones they grappled with at the height of civil strife from 1978 to 1983. What sorts of items, of late, constitute the daily fare of the Guatemalan press but seldom register here in Canada? A memorable trip in the summer of

2013 – I have made over sixty in the past forty years – affords me the opportunity to peruse my press clippings and relay what follows.

A Diminished Army, but One That Still Has Clout

The day before Canada Day is Army Day in Guatemala. I arrive in the late afternoon and pick up the *Prensa Libre*, the country's most widely read newspaper, which I read in the peace and quiet of Antigua, the old colonial capital I much prefer to the bedlam of Guatemala City. Hitherto a gala spectacle of medalled authority and military paraphernalia, Army Day isn't what it used to be, primarily because the national armed forces aren't what they used to be. "An Army of Leftovers" declares the headline in the *Prensa Libre* of 30 June.[1] A four-page feature documents the military's erosion, the result of the implementation of the Peace Accords, which called for a significant reduction in army personnel and its prominent role in everyday life. Today the ranks of the army total some 23,000, two-thirds its size at the pinnacle of counter-insurgency in the early 1980s, when attacks on unarmed civilians ushered in a reign of terror and, in the eyes of a UN Truth Commission, constituted acts of genocide when perpetrated against Indigenous Maya communities in remote highland regions.[2] Now with one soldier for every thousand inhabitants, Guatemala's is the lowest such ratio in all Latin America. The decline in manpower is paralleled by a chronic lack of equipment, leading *Prensa Libre* to conclude that "a weak institution with paltry resources is unable to guard even minimally, by land, sea, or air, our nation's borders." Have the mighty fallen that far?

Statistics invariably conceal as much as they reveal. The former president of Guatemala Otto Pérez Molina is himself a retired general. Under his watch the army took to the streets and penetrated the countryside, purportedly to increase public security in the face of widespread gang violence and rampant drug-trafficking. The very next day, a photograph of Pérez Molina in the *Prensa Libre* shows him, as commander-in-chief of the national armed forces, proudly surveying his troops. Mention is made of the presence in Guatemala of U.S. Secretary of State John Kerry, there to mark the delivery from the U.S. Southern Command of forty-two military vehicles along with the counsel of U.S. military advisers. Given the impact of past U.S. interference in Guatemalan affairs – the CIA was behind the overthrow in 1954 of democratically elected president Jacobo Arbenz Guzmán, whose program of moderate agrarian reform was deemed a threat to American business interests, those of the United Fruit Company in particular – Mr Kerry cast an ominous spell simply by ghosting in and out around Army Day.

According to analyst Jeff Abbott, after Pérez Molina assumed the presidency in 2012 "the amount of military aid provided to Guatemala by the US government, despite the limits placed by Congress, increased nearly 40 percent," with more than "30 public ministries headed by former military." Conspicuous military involvement, Abbott asserts, signals "a return of the strategy of counter-insurgency, and of corruption" (2015). One of his sources, Kelsey Alford-Jones of the Guatemalan Human Rights Commission in Washington, DC, further contends that the re-militarization of public life "endangers those who are fleeing violence or persecution and has done almost nothing to address entrenched corruption, trafficking, and deadly violence."[3]

As his term in office wound down, Pérez Molina became a president besieged. Corruption scandals and allegations that drug money lay behind the financing of political parties reached a crescendo in the months leading up to the September 2015 elections. Pérez Molina's vice-president, Roxana Baldetti, was forced to resign from office in the wake of revelations of massive fraud in customs operations, in which one of her aides, Juan Carlos Monzón Rojas, was implicated in a scam known as *"la línea."* In the words of an Associated Press release of 9 May 2015, officials caught up in *la linea* "defrauded the state of millions of dollars by taking bribes to lower customs duties." Monzón, after five months on the run, turned himself in on 5 October. After news broke of this particular scandal, peaceful weekly demonstrations, an unusual occurrence in Guatemala, saw thousands gather and call for Pérez Molina to follow his vice-president's example and tender his resignation also. This he eventually did, on 2 September, whereupon (following a court hearing about his alleged role in *la linea* and only four days before the first round of elections) he was subsequently imprisoned, as Baldetti had been earlier. Pérez Molina, like Baldetti, remains in captivity awaiting trial on multiple charges of wrongdoing, charges that both deny. A second round of elections, held on 25 October, was won by Jimmy Morales, a popular TV comedian who surprised most observers by the margin of his victory. No laughing matter, Morales heads a party staffed and supported by retired members of the military high command.

The Guatemalan army may be diminished in number, but it still yields considerable clout.

Malnutrition in a Bountiful Land

Statistics once again suffuse the pages of *Prensa Libre*, those in the issue of 2 July 2013 disconcerting in the extreme. "Supplies of Corn and Beans Are Running Out," the headline states, food security as opposed to public security being undermined on account of "high prices and low purchasing power," with "poor families in dire need." Amid reports of malnutrition on par with African levels plaguing eastern parts of the country, a survey by the UN Food and Agriculture Organization (FAO) identifies issues of access, not overall production, as the crux of the problem. Its incisive analysis is echoed by the findings of local officials. Despite the juggernaut of globalization that has transformed age-old ways, Guatemala is still at heart an agrarian society. Nor is it a poor country; it has been made so by crippling geographies of inequality, especially in land tenure. Some 65 per cent of agricultural holdings are controlled and operated by 3 per cent of owners, a consequence of which is that a mere 15 per cent of total farmland is shared among more than 85 per cent of all other owners. Most rural Guatemalans can no longer feed themselves adequately from their own plots; the native peoples epitomized by Miguel Ángel Asturias (1899–1974) in his novel *Hombres de maíz* ([1949] 1972; *Men of Maize*, 1975) as being precisely that, have long served as hired hands for wealthy plantation owners who grow crops for export, not internal consumption. The FAO survey ascertains that the monthly wage for agricultural labourers averages 2,421.75 quetzales (roughly US$300), falling short of the cost of a "basic minimum" to feed a family by 350.25 quetzales. Guatemala has not yet, in the fitting words of historian Severo Martínez Peláez (1925–98), made the leap "desde una patria de pocos hacia una patria de todos" (Martínez Peláez [1970] 1998, 524) meaning gone from a homeland for a few to a homeland for all). Any talk of agrarian reform is frowned upon, and nixed, by the elite lobby known as the Comité Coordinador de Asociaciones Agrícolas, Comerciales, Industriales y Financieras (CACIF; the Coordinating Committee of Agricultural, Commercial, Industrial and Financial Associations), whose priorities dominate and prevail, straitjacketing the political economy of Guatemala and castigating any group or individual perceived to challenge the status quo.[4]

Genocide, Justice, and the Trial of Ríos Montt

The *Prensa Libre* on 3 and 4 July 2013 buries in interior pages the state of play in the trial of another former president (and army general) Efraín Ríos Montt. He stands accused, along with the former head of military intelligence, José Rodríguez Sánchez, of genocide and crimes against humanity, specifically in relation to Ixil Maya communities in 1982 and 1983, when his seventeen-month presidency witnessed among the worst of atrocities committed during the armed conflict. On 28 January 2013, Judge Miguel Ángel Gálvez ruled that the case against Ríos Montt would be heard in Guatemala's Supreme Court, the first such tribunal anywhere in the world in which a one-time head of state was placed on trial for genocide in the country he governed. On 10 May, presiding judge Yassmin Barrios and her two associates, Patricia Bustamante García and Pablo Xitumul de Paz, found Ríos Montt guilty and sentenced him to eighty years in prison, where he was dispatched directly from the courtroom. Elite outrage was articulated by CACIF's vehement criticism of the verdict, which saw it denounced in a series of inflammatory pages paid for in newspaper campaigns. While the deeds deliberated upon are generally agreed to have occurred, the charge of genocide is disputed and denied.[5] Furthermore, Ríos Montt and his lawyers assert that he has immunity from prosecution on the basis of an amnesty granted by the military dictator who ousted him from power, General Humberto Mejía Víctores. For ten days, justice appeared to have been served. Impunity in Guatemala, however, runs deep; its beneficiaries made it clear that the Supreme Court's ruling would not be tolerated. On 20 May, citing lack of due legal diligence, the Constitutional Court overturned the verdict and ordered a retrial, a charade that began in January 2015. Ríos Montt, frail of health but defiant of spirit, was wheeled into the courtroom on a stretcher, only for the tribunal to be adjourned, reconvened, and adjourned once again, pending medical reports that the defendant be certified fit to stand trial. Death in legal limbo, aged ninety-one, was Ríos Montt's ultimate fate. His daughter, Zury Ríos Sosa, aspired to run for the presidency in the September 2015 elections, her desire to do so in contravention of a law that prohibits relatives of former dictators running for executive office. Like father, one wonders, like daughter?

Homicide, Public Security, and Gang Culture

The Fundación para el Desarrollo de Guatemala (FUNDESA; Foundation for the Development of Guatemala) does all it can to attract investors, but the enterprise admits that its own registers of public security (or lack thereof) make the country a hard sell. In 1995, before the signing of the Peace Accords, FUNDESA recorded the homicide rate in Guatemala at 33.8 for every 100,000 inhabitants. It currently estimates the figure at 32.3 (Canada's in 2011 was 1.62, a forty-four-year low). In 2006, the tenth anniversary of the Peace Accords, the index was actually higher than the pre-Accords figure: 46.3 murders for every 100,000 inhabitants. Most violent deaths are never investigated, let alone brought for due process before the courts.

Mauricio López Bonilla, minister of the interior, explains in the *Prensa Libre* of 5 July 2013 that the period 2004–8, under the presidency of Oscar Berger, saw not only the army reduced in size but the police force too, the latter's "less professional training" and its loss of "civil intelligence staffing" compromising its operations further. The army and the police, López Bonilla implies, are no match for organized crime, be it drug cartels from Mexico that orchestrate the transfer of cocaine from Colombia to the United States or the local gangs, known as *maras*, recruited for a range of middleman activities, from extortion and intimidation to ensuring safe passage and settling scores. Canadian Adam Blackwell, the newspaper *El Periódico* of 5 July reports, is the secretary of a branch of the OAS called "Multidimensional Security." Amidst talk of decriminalizing the drug business writ large – Guatemalans and other Latin Americans, Pérez Molina to his credit once pointed out, are the ones who pay the highest price for what is overwhelmingly a U.S. habit – Blackwell's job entails decriminalizing one aspect of it, the existence and role of the *maras*. He is praised for brokering deals in neighbouring El Salvador and Honduras that have seen rival gangs in both countries cease animosities among themselves and enter a pact agreed to by government that seeks to integrate the delinquents back into society. "Without talking with them," Blackwell states firmly, "it will be very difficult to lower the number of homicides in Guatemala," which he puts at twice that of FUNDESA, a chilling 95 for every 100,000 inhabitants. "We are still unclear," he adds, "whether or not the government of Guatemala wishes to embark on a process similar to that of El Salvador and Honduras. The messages that I am getting from gang leaders is that a deal might be possible." After Blackwell's negotiations in El Salvador, homicides there, according to the national police, dropped by one-third, from fifteen a day to five or six. Fatal acts of violence and concerns for public security, however, persist. Indeed, homicide rates have risen to unprecedented levels since the pact between gangs and government, and between rival gangs, which lasted from March 2012 to March 2014, came to an end. In June 2015, 677 homicides were recorded in El Salvador, an average of more than 22 a day.

Mara omnipresence and the grip of gang culture are often cited by young people, and their parents, as the reason behind the decision to flee embattled homelands and seek a better life in the United States. In 2014, some 68,500 minors were documented as having made the perilous trek to El Norte, many of them doing so with parental sanction and financial assistance: their sons and daughters, parents reckon, have a better chance of making it to adulthood if they can get out of Central America. The phenomenon prompted President Obama to allocate $1 billion to combat the mix of factors that propels the exodus of youth from the "northern triangle" of Guatemala, Honduras, and El Salvador. Despite Obama's efforts, the flow continues: the *Prensa Libre* of 8 April 2015 reported detentions by "migration authorities" for the first quarter of the year at "15,647 unaccompanied minors," of whom 5,465 are registered as Guatemalan, 2,788 Salvadoran, and 1,549 Honduran, with citizens of Mexico in the majority. Deportations are also at record levels, Mexican law enforcement officers as dutiful in this regard as their American counterparts, though the latter have the statistical edge by a considerable margin.[6] Upon arrival back home, however, many deportees simply turn around and retrace their steps, parents and their offspring undeterred by horrific episodes that have seen youthful migrants apprehended en route through Mexico and pressed into service as drug mules, or have their families extorted for their safe release. If deemed of no mercenary or monetary worth, the most luckless are killed or left to die, abandoned in a borderland desert where vigilante justice can be as lethal as wrathful natural elements.[7]

Crossing the Isthmus

Geography is fundamentally what Central America is about. Just as the isthmus today is a corridor that connects areas of high demand for drugs with regions of ready supply, so too, historically, is what Pablo Neruda in one of his poems called "la dulce cintura de América" (Neruda [1950] 1998, 335; the sweet waist of America), a narrow strip of land that separates the world's two great oceans. Finding ways to link the Atlantic and the Pacific, and facilitate trade between them, has been the goal of fortune seekers ever since Vasco Núñez de Balboa (not "stout Cortez," as John Keats erroneously put it) stood "silent, upon a peak in Darien" five centuries ago, the first European alleged to have trekked from the Atlantic littoral to contemplate the Pacific. No one has captured a sense of Central America's geopolitical destiny better than William Paterson, the financier whose advocacy of the isthmus as a strategic hub during Scotland's disastrous attempts to colonize the Darien region of Panama in the late seventeenth century is enshrined in his describing it as "Door of the seas, key of the universe."[8] Up until 1890, Nicaragua was the preferred choice of the United States as the country through which to finance the construction of an interoceanic canal. Panama won out in the end; since 15 August 1914, each day thirty ships or more have plied its eighty-kilometre-long double set of locks, built at a cost of $639 million and 20,000 working lives. To accommodate an ever-increasing volume of traffic – some 14,000 freighters passed through the Panama Canal in 2012 – a third set of locks has now been constructed. The *Prensa Libre* of 7 July 2013, however, notes that Nicaragua, with Chinese backing, is again promoting its territory as an alternative route.[9] A joint venture is also championed between El Salvador and Honduras. Guatemala has its own ambitious project, a "dry canal" constituting two highways and a railroad between Puerto Barrios on the Atlantic and Puerto San Luis on the Pacific, the cost of traversing 372 kilometres of terrain estimated at some $5 billion.

Mining, Development, and Canadian Complicity

On two consecutive days, 12 and 13 July 2013, *Prensa Libre* devoted prime coverage to the aftermath of a television broadcast made by then-president Pérez Molina on 9 July, in which he proposed that a moratorium be placed on the granting of new licences for mining. The business elite, predictably, oppose such a proposal, and took a dim view of Pérez Molina's pointed words about its attitudes and behaviour. Environmentalist groups like the Madreselva Ecological Collective call either for a total suspension or genuine, meaningful consultation before mining operations begin (see chapter 4 by Magalí Rey Rosa). Guatemala has over 100 mines in production, with applications to open some 350 more. Foreign companies are lured to Guatemala because of favourable concessions granted them that, in effect, flout environmental and legal legislation, generating intense conflict in communities whose lands lie closest to, or indeed form part of, extractive operations carried out by local subsidiaries. Canadian companies are in the forefront of investment. In the words of Alain Deneault and William Sacher, "Canada stands out as a judicial and financial haven that shelters its mining industry from the political or legal consequences of its extraterritorial activities by providing a lax domestic regulatory structure that it seeks to export through international agencies, diplomatic channels, and 'economic development projects.'"[10]

Not, however, at least by one ruling, for much longer. Far from the scene of turmoil that saw murder and rape take place at its former site in El Estor, Toronto-based Hudbay Minerals was summoned by Ontario Superior Court Justice Carole Brown to stand trial in Canada for alleged abuses of human rights committed by its affiliates, HMI Nickel Inc. and the Compañía Guatemalteca de Niquel (CGN), in Guatemala in 2007 and 2009 (see chapter 3 by Kalowatie Deonandan and Rebecca Tatham). "Will Canadian Companies Be Held Responsible for Their Actions Abroad?" asked an op-ed piece by Shin Imai in the *Globe and Mail* of 31 July 2013 after the landmark ruling of Justice Brown. The question hovers, but a precedent has been set, one that reverberates in the hitherto cocoon realm of corporate Canada. Murray Klippenstein, a lawyer representing Indigenous Q'eqchi'-Mayas whose lives and livelihoods have been adversely affected by Hudbay's actions, put it succinctly in a press advisory prepared by his law firm: "There will now be a trial regarding the abuses that were committed in Guatemala, and this trial will be in a courtroom in Canada, a few blocks from Hudbay's headquarters, exactly where it belongs." Mr Klippenstein adds, "We would never tolerate these abuses in Canada, and Canadian companies should not be able to take advantage of broken-down or extremely weak legal systems in other countries to get away with them there" (Klippensteins Barristers and Solicitors 2013).[11]

My visit to Guatemala in 2013 ended with my attending a conference at which a forum on Canadian mining activities sparked lively debate. During her presentation, one of my graduate students, reporting on the findings of dogged fieldwork, informed the audience that a new Canadian company recently awarded a licence is called Gunpoint Exploration. *"Nomina sunt consequentia rerum* [Names are the consequences of things]" runs a line in Dante's *La Vita Nuova* (Bryars 1994, liner note).

The poet has spoken.[12]

NOTES

1 Translations from articles in *Prensa Libre* are by the author.
2 For a reconstruction of the armed conflict in Guatemala, like this contribution based on a reading of Guatemalan newspapers, see Lovell ([1995] 2010,

48–104). For a discussion of the operation and findings of the UN Truth Commission, see Lovell ([1995] 2010, 158–62). Also see the full text of the Conclusions of the CEH report, *Guatemala: Memory of Silence* (2009). With customary erudition, to say nothing of quirky prowess, Nelson (2015) subjects the charge of genocide, refuted still by many powerful sectors of Guatemala society, to moral as much as mathematical scrutiny.

3 See Abbott (2015) for elaboration. He points out that increased militarization in the public sphere "is a direct violation of the 1996 Peace Accords. Specifically, it violates the mandate for division of the police and the military. And this has occurred with the direct complicity and support of the United States government," which continued to relay "unflinching support" for Pérez Molina until just before his precipitous fall from grace.

4 In a visceral op-ed piece published in *El Periódico* on 15 August 1999, the Guatemalan writer Mario Monteforte Toledo had this to say about the land question:

> Land is the root cause of national backwardness, of elite economic clout, of social imbalance, of the survival of pre-capitalist structures, of the overpopulation of our cities, of criminality, of the absence of internal markets, of menacing unrest in rural areas, of ignorance, of illiteracy, of the nostalgia that was this country five centuries ago.
>
> A million or so peasants who know no other toil than to work the land, have none; two and a half million more scratch away at miserable plots on the sides of mountains and lay waste to forests in order to supplement their starvation wages; three hundred thousand others leave to work in Mexico each year between the months of October and January; half a million are bought and sold, as if they were cattle, by labour contractors who deploy them here in Guatemala on coffee and sugar plantations while their corn grows in the highlands; more than half a million, for years now, possess pieces of paper that entitle them to land, but at the same time there are more than ten thousand disputes among their communities over the rights to ownership, for which they kill one another; in the plantation zones, hundreds of peasants are murdered because they are thought to be "dangerous" or are believed to be "terrorists"; hundreds more plantation owners feel threatened by discontented Indians and so arm militias in order to defend their properties and themselves; and the countryside has given birth to its own form of justice [in the form of mob lynchings], because the state simply does not exist there and the rule of law does not apply.
>
> Of the inclination of political parties and presidential candidates not to engage the land question, he concludes, "There appears to be a

consensus not even to raise the matter of the most flagrant deformity in our country. The only explanation I can offer for this act of concealment is the fear of sounding 'like a communist' – and so not compromise oneself with respect to solutions should one's party afterwards become the government" (translation by the author).

5 The Truth Commission (Lovell [1995] 2010, 161–2) is very precise in its declaration of genocide, wording its statement under the terms of reference of the Geneva Conventions. Part of the declaration runs:

The [Truth Commission] concludes that agents of the State of Guatemala, within the framework of counterinsurgency operations carried out between 1981 and 1983, committed acts of genocide against groups of Maya people who lived in the four regions analyzed. This conclusion is based on the evidence that, in light of Article II of the Convention on the Prevention and Punishment of the Crime of Genocide, the killing of members of Maya groups occurred (Article IIa), serious bodily or mental harm was inflicted (Article IIb), and the group was deliberately subjected to living conditions calculated to bring about its physical destruction in whole or in part (Article IIIc). The conclusion is also based on the evidence that all these acts were committed "with intent to destroy, in whole or in part," groups identified by their common ethnicity, by reason thereof, whatever the cause, motive, or final objective of these acts may have been.

6 In a sobering analysis, Price and Breese (2016, 369) record that 4.2 million people were deported from the United States between 2000 and 2013, a tally in which "Latin American countries accounted for 94 percent of all removals." Not surprisingly, "Mexico dominates, accounting for seven out of every ten removals," a total of some three million, with Central America logged at almost 800,000. For the latter, the "incidence of removal" shows a dramatic surge after 2006, with Guatemala pegged at 296,110, Honduras at 275,113, and El Salvador at 183,418.

7 The plight of migrant youth is nowhere more poignantly portrayed than in two feature films, *Sin Nombre* (2009, directed by Cary Joji Fukunaga) and *La Jaula de Oro* (2013, directed by Diego Quemada-Diez). The former tells the story of Willy, better known by his gang name of El Caspar, and the innocent but quick-thinking Sayra, their situation as they venture north made even more hazardous by the fact that El Caspar is fleeing the retribution of his former associates in the Mara Salvatrucha, perhaps the most

notorious gang in the region. In the latter, a resourceful threesome is whit-
tled down to one sole survivor. His job, once he reaches the promised land,
is to mop up in an American slaughterhouse, a grisly but apt metaphor for
the migrant collective lot. Both films involve harrowing scenes aboard or
along the trajectory of *La bestia* (The beast) – the name given the train that
winds its way from Mexico's southern border with Guatemala to its northern limit with the United States. Stephanie Nolen (2015a, 2015b) contextu-
alizes the *mara*-migrant dynamic in El Salvador, and the grim "post-pact"
homicide statistics that afflict the country, in two superb pieces of intrepid
investigative reporting.

 8 Paterson, cited in Prebble ([1968] 1970, 4). See also Lovell (2008) and Lovell
   and Lutz (2011) for further discussion of Darién and the Central American
   isthmus in the global scheme of empire.

 9 With characteristic aplomb, Anderson (2014) discusses President Daniel
   Ortega's intent, against massive opposition that unites Nicaraguans of all
   classes and ideologies, to construct what the *New Yorker* staff writer calls
   "The Comandante's Canal."

10 Deneault and Sacher (2013, 2). Produced and directed by Rachel Schmidt,
   the film *Defensora* (2013) documents how the operations of one Canadian
   mining company and its Guatemalan affiliate have wrought such trauma
   and negative impact on the Q'eqchi'- Maya of El Estor (see Candace Johnson in chapter 1 and Catherine Nolin in chapter 2).

11 In May 2015, three Maya plaintiffs, Angélica Choc, Germán Chub, and
   Rosa Ich, travelled to Canada to present their case not in a court of law
   but to shareholders gathered in Toronto for the annual general meeting of
   Hudbay Minerals. Another thirteen Maya plaintiffs presented testimony
   to Hudbay lawyers in Toronto in November 2017, called upon to do so in
   connection with a precedent-setting lawsuit.

12 An earlier version of this chapter was published as "Postcard from Guatemala," *Queen's Quarterly* 121, no. 4 (2014): 567–81.

WORKS CONSULTED

Abbott, Jeff. 2015. "U.S. Policy Driving Militarization in Guatemala." *Truthout*
   5 July.
Anderson, Jon Lee. 2014. "The Comandante's Canal." *New Yorker*, 10 March,
   50–61.
Asturias, Miguel Ángel. (1949) 1972. *Hombres de maíz*. Madrid: Alianza
   Editorial.

– 1975. *Men of Maize*. Translated by Gerald Martin. New York: Delacorte.

Alighieri, Dante. (1295) 1973. *La Vita Nuova*. Translated by Mark Musa. Bloomington: Indiana University Press.

Bryars, Gavin. 1994. *Vita Nova*. ECM Records, new series 445 351-2.

Commission for Historical Clarification. 2009. *Guatemala: Memory of Silence. Report of the Commission for Historical Clarification: Conclusions and Recommendations*. https://www.aaas.org/sites/default/files/migrate/uploads/mos_en.pdf.

Deneault, Alain, and William Sacher. 2012. *Imperial Canada Inc.: Legal Haven of Choice for the World's Mining Industries*. Vancouver: Talon Books.

Imai, Shin. "Will Canadian Companies Be Held Responsible for Their Actions Abroad?" *Globe and Mail*, 31 July 2013. https://www.theglobeandmail.com/opinion/will-canadian-companies-be-held-responsible-for-their-actions-abroad/article13524877/.

Joji Fukunaga, Cary, dir. 2009. *Sin Nombre*. With Paulina Gaitán, Marco Antonio Aguirre, Leonardo Alonso. Scion Films.

Klippensteins Barristers & Solicitors. 2013. "Press Advisory: Ontario Court Rules That Lawsuits against Hudbay Minerals Regarding Shootings, Murder, and Gang-Rape at Its Former Mine in Guatemala Can Proceed to Trial in Canada." http://www.chocversushudbay.com/wp-content/uploads/2010/10/PRESS-ADVISORY-Ontario-Court-rules-claim-can-proceed-to-trial.pdf.

Lovell, W. George. (1995) 2010. *A Beauty That Hurts: Life and Death in Guatemala*. 2nd rev. ed. Austin: University of Texas Press.

– 2008. "The Darien Gap." In *Encyclopedia of Latin American History and Culture*, ed. Jay Kinsbruner, 2nd ed., 2:746–7. New York: Charles Scribner's and Sons.

Lovell, W. George, and Christopher H. Lutz. 2011. "Between Two Seas: Antonio de Herrera and the Mapping of Central America." In *Mapping Latin America*, ed. Jordana Dym and Karl H. Offen, 65–9. Chicago: University of Chicago Press.

Martínez Peláez, Severo. (1970) 1998. *La patria del criollo: Ensayo de interpretación de la realidad colonial guatemalteca*. Mexico City: Fondo de Cultura Económica.

Monteforte Toledo, Mario. 1999. "Op-ed Commentary." *El Periódico*, 15 August.

Nelson, Diane M. 2015. *Who Counts? The Mathematics of Death and Life after Genocide*. Durham, NC: Duke University Press. https://doi.org/10.1215/9780822375074.

Neruda, Pablo. (1950) 1998. *Canto General*. Edited by Enrico Mario Santí. Madrid: Cátedra.

Nolen, Stephanie. 2015a. "Menaced by Gangs, El Salvador's Children Are Running for Their Lives." *Globe and Mail*, 29 August. https://www.theglobeandmail.com/news/world/menaced-by-gangs-el-salvadors-children-are-running-for-their-lives/article26151568/.

– 2015b. "Under Siege." *Globe and Mail*, 21 August. https://www.theglobeandmail.com/news/world/under-siege-el-salvadors-state-ofchaos/article26054872/.

Prebble, John. (1968) 1970. *The Darien Disaster*. Harmondsworth: Penguin Books.

Price, Marie, and Derek Breese. 2016. "Unintended Return: U.S. Deportations and the Fractious Politics of Mobility for Latinos." *Annals of the Association of American Geographers* 106 (2): 366–76.

Quemada-Diez, Diego, dir. 2013. *La Jaula de Oro / The Golden Dream*. With Brandon López, Rodolfo Domínguez, Karen Martínez. Animal de Luz Films.

Schmidt, Rachel, dir. 2013. *Defensora*. Of Earth and Sky Film.

10 Conclusion

STEPHEN HENIGHAN AND CANDACE JOHNSON

On 29 December 2016, Guatemala celebrated the twentieth anniversary of the signing of the Peace Accords that ended a thirty-six-year civil war. The formal celebrations, held in the National Palace, were directed by President Jimmy Morales, who was flanked by former presidents Alvaro Arzú (1996–2000) and Marco Vinicio Cerezo Arévalo (1986–91). President Morales spoke in philosophical terms about peace and cited Jesus Christ and Nelson Mandela in order to emphasize the magnitude of the commitment to peace as an aspiration for the country as it continues its struggle for justice. His words were hopeful as he referred to the internal conflict and ethnic divisions as problems of the past, barely acknowledging the persistence of problems that initiated and fuelled the war in the first place – internal and external colonialism, marginalization of Indigenous communities, racism, and state-sponsored violence. Quite predictably, there was no mention of a genocide. Rather, his concluding message to the world was that Guatemala is a model of what is possible: "Hoy Guatemala les recuerda al mundo que la paz sí es posible, que la tolerancia y el respeto a las ideas también es posible. Eso gracias a quienes nos heredan la paz, representados a través de la firma de una paz firme y duradera. Ahora nos corresponde a todos nosotros a dar pasos firmes a favor de las nuevas generaciones para una paz duradera" (Morales 2016; Today Guatemala reminds the world that peace is possible, that tolerance and respect for ideas is also possible. This is thanks to those who provide us with the legacy of peace, represented by the signing of a strong and lasting peace. Now it is up to all of us to take steps to ensure a lasting peace for new generations). He spoke of the "nueva generación" earlier in his speech as well, asking what the new generation will do with the peace that they have inherited. Employing the words of Nelson Mandela, he reminded them that "Todo parece imposible, hasta que se hace" (Everything seems impossible, until it is done). This is hopeful and ambitious and effectively captures the spirit that will be necessary for future work towards peace, justice, and democracy. But is it true? Is tenacity the antidote to the impossible? The posters designed by the Morales government to mark the twentieth anniversary of the Peace Accords, which appeared on billboards along highways, and at the stations of Guatemala City's new TransMetro rapid bus system, suggest that the country's current rulers have little dedication to solving Guatemala's long-term culture of injustice. Emblazoned with the slogan "Las guerras son inútiles" (Wars are useless), the posters were interpreted by many as dismissing the sacrifices made by those who died in the effort to forge a just society; some read the posters as a swaggering boast by the elite that not even the most determined struggle would dislodge them from their positions of privilege.

This book documents the structural injustice that persists in Guatemala and its impact on individuals, communities, and the environment. The story of institutions – the army, the government, neocolonial corporations, patriarchy – raises serious questions about the prospects for justice in post–Peace Accords Guatemala. The evidence provided in each of the chapters makes a case for the unlikelihood of progress towards a just society, even in the context of transnational support, including institutions such as CICIG, international pressure exerted by NGOs, and mass protest. The impressive efforts to achieve transitional justice by testing the state's capacity to answer to citizens' demands has resulted in small but significant victories, moments of justice that have transformational potential: the prosecution of Ríos Montt for genocide; the prosecution of the police officers who killed the detective who was working on the Myrna Mack case; the prosecution of Roxana Baldetti and Otto Pérez Molina for corruption; the criminal case against Mynor Padilla for mining-related violence, and the civil cases against Hudbay for the same incidents in the Canadian courts; the 2016 victory of the women in the Sepur Zarco trial, who testified to the horrifying sexual abuse that they suffered at the hands of the army in the 1980s. They endured and won. These are the *granitos* of progress and hope that are painstakingly collected in efforts to tip the scales of justice towards fairness, reconciliation, and democracy.

These achievements, however, must be measured against continued evidence of structural weakness. In January 2017, President Morales's brother and son were detained by police for alleged corruption and tax fraud (*Wall Street Journal* 2017). This family hardship was likely the reason for the president's announcement later that month that he would no longer be able to donate 60 per cent of his salary to important causes (Martín 2017). In addition, on 8 February 2017, the Associated Press reported, "A Supreme Court justice has been arrested on a charge of influence peddling for trying to help her son in a corruption case." And these are not isolated incidents. The headlines documented by George Lovell (chapter 9) reveal deep contradictions in Guatemalan society and politics and the absurdity of a country that is simultaneously highly self-aware and resolutely ignorant. Further, these events serve as reminders of the culture of impunity that pervades Guatemala and exists in stark contrast to the reality of life for most people who live in the country. As explained in the introduction to this volume, distributional inequity is linked to the project of transitional justice and is just as resistant to change. Poverty is the status quo for the majority of the population, which is exacerbated by chronic droughts and the Zika virus (among other things), both likely related to climate change.

The transnational nature of injustice is also apparent in Guatemala. Climate change is borderless, yet much of the burden is shouldered by the countries of the Global South. The yearly droughts create increased food insecurity and starvation for people living in Central America's dry corridor while Guatemala's agro-food export industry thrives. Canadian mining companies continue to operate in Guatemala, benefiting from weak institutions and lax environmental standards, while providing little for the communities that are most affected by their activities. However, as Magalí Rey Rosa makes clear in chapter 4, mining is a dubious yet complex business in Guatemala. The results are predictable, but the wider context is beginning to shift. Deonandan and Tatham (chapter 3) examine the intricacies of the Padilla case, connected to harms caused by Hudbay, and Nolin (chapter 2) connects mining struggles to broader human rights issues. While none of these cases provides evidence of the goodwill of the Canadian government at any time or of any partisan stripe, Canadian courts have provided avenues for redress. In 2013, a court in Ontario ruled that the plaintiffs who filed cases against Hudbay for harms committed in Guatemala should have their cases heard in Canadian courts and that Hudbay could be held responsible for damages. This was precedent setting, as it was the first time that a Canadian court would hear cases concerning the conduct of Canadian corporations abroad. Similarly, in 2017,

British Columbia's Court of Appeal decided that Guatemalan plaintiffs in another case (against Vancouver-based Tahoe Resources) would be granted standing in Canadian courts (Bernard 2017; Taylor 2017). These cases, along with another decided in BC concerning Eritreans "alleging human rights abuses while working at a Vancouver-based mine in their home country" (Bernard 2017), demonstrate a trend towards greater responsibility for justice across borders.

This is a significant development and one that is likely to expand possibilities for redress and the recognition of global linkages and obligations. Francisco Goldman, quoting Michel Foucault, claims in esoteric terms that peace might be merely war that is waged by the state (Goldman 2007, 63). If this is the case, then it might be only external venues and forces that can create the sort of peace that is not continued war, but rather a path to justice. However, these prospects for justice run parallel to Guatemala's national struggle for transitional justice, and civil cases abroad can address only individual, particular harms and not those committed against entire communities and across generations. Further, the distributional benefits that might accrue from international legal action are likely to be minor, as distributional equity in the form of social services and access to responsive state institutions is the responsibility of the nation state itself. Therefore, the potential for transnational justice might have little currency, on its own, for most Guatemalans.

Such is the dismal, yet not hopeless state of structural injustice in Guatemala. But what about individual agency? How do individuals and groups circumvent, resist, and negotiate with oppressive structural constraints? This book also documents the action and analysis of key political actors and human rights activists: environmental activist and journalist Magalí Rey Rosa (chapter 4), human rights activist Helen Mack Chang (chapter 5), and former attorney general Claudia Paz y Paz (chapter 6). These three women have documented human rights and environmental abuses and demanded governments be held to account for the persistent violations, brought about the convictions of the material authors of murder in the case of Myrna Mack, and prosecuted a former president for genocide, respectively. And there are many more women and men working against futility, risking everything for justice. In fact, this is what is unique about Guatemala (although it might not be unique *to* Guatemala): in a country that bears the burden of a colonial legacy, a recent genocide, military dictatorship that masquerades as democracy, endemic corruption and political scandal, as well as poverty and inequality, individuals conceive of and advance ambitious projects to insist on respect for human rights and advance justice.

The importance of individuals in the struggle for justice is also revealed by the work of the artists that are the subjects of chapters 7 and 8. Rita Palacios's powerful description and analysis of two of Guatemala's contemporary artists make evident the significance of cultural critique. Palacios explains that in the performance art of Regina José Galindo, "the body is another site where memories reside and where they can be deployed.... In [Galindo's] performances, audiences are confronted with a female body that carries with it the memory of a violence that take them back to the former, and in many cases to the current, Guatemalan state." This description reinforces the need to understand Guatemala's struggle for justice in the context of the "present past" (Oettler 2006, 3–4), as transitional justice suffers from a temporal amalgamation; what is in the past is also in the present and cannot be separated from corporeal and metaphysical reality. Similarly, Palacios identifies Rosa Chávez as "a Quiché-Kaqchiquel-Maya woman, [who] writes a poetry that is performative and that permits the writer and her reader to engage with notions of gender and ethnicity, as she sets out to defy her reader with a body that is sensual, sexual, female, and Indigenous" (Palacios, this volume, 152). The work of these artists demonstrates the constrained capacities for human action and expression within violent environments that are ambivalently attuned to that violence and desensitized to it.

The contradiction of political awakening is also one theme of the life and work of Rodrigo Rey Rosa, examined by Stephen Henighan in chapter 8. Rey Rosa's journey avoided political terrain until his literary career had been well established, and his turn to the political was not a testimony of personal suffering – though his mother's kidnapping in 1981 altered his perception of Guatemalan reality – but the expression of environmental concerns that implicated social and political realms. Similar to the work of Galindo and Chávez, the individual artist, or work of art, is a conduit for expression of truths about a violent, fragile environment. Regardless of whether this environment is understood in primarily political or physical terms, the subjects of the art tell truths about that environment and demonstrate the consequences of ignoring the human voices. Further, Rey Rosa's work confirms the dissonance of time, as the writer's own political consciousness, reflected through his characters, is located in the "present past." Henighan explains, "Near the conclusion of Los Sordos, Clara Casares and her lover Javier discuss whether a writer acquaintance named Rodrigo can be considered 'producto de su tiempo y su medio" (a product of his time and his milieu), with Clara expressing the forceful opinion that the writer does not belong to his social class: 'Ése está, y siempre estuvo, fuera del tiempo – agregó, pero en un tono despectivo, descalificador – y en ningún lugar, si me preguntas a mí'" (Rey Rosa 2012, 215; "That guy is, and always was, outside time," she added, but in a disdainful, disqualifying tone, "and nowhere at all, if you ask me").

Where does that leave us in terms of progress towards justice? First, it leads to the conclusion that time is not itself an indicator of progress. Twenty years of peace does not tell us much about what has been achieved and what is left to pursue. The individual chapters in this volume explain the complexities of transitional and post-transitional justice and reach different conclusions about structural injustice and the possibilities and limitations of individual and collective action. However, the contributors are consistent in their acknowledgment of limited yet significant gains in the face of considerable challenges. None is hopeless or dismissive. The differences in analysis are the products of different cases, circumstances, evidence, and standpoints. It is worth noting that those who have focused primarily on structural violence and the institutions of the state – Rey Rosa, Mack Chang, Paz y Paz, Deonandan and Tatham – are marginally more hopeful (despite having direct experience with the gravest consequences of the violence generated by state institutions). Those that focus on cultural violence, to use Matthew Mullen's distinction (2015, 463) – Nolin, Henighan, Palacios, and Lovell – are much less sanguine. This suggests that the cultural elements of violence are the most recalcitrant (ibid.) and the least amenable to change through international declarations and institutional reform.

This disheartening truth is captured well in Jayro Bustamante's film *Ixcanul* (2015), the first Guatemalan film to be nominated for an Academy Award. The film depicts life in an Indigenous community on a coffee plantation on Guatemala's Pacific slope. The family at the centre of the story is marginalized in multiple ways. They are poor. They depend on the plantation's foreman for their continued employment. Their only child is a girl (and therefore a financial and familial burden), whom they have arranged to marry to the relatively affluent and powerful plantation foreman. They are beholden to traditional beliefs, in spite of their awareness of their inefficacy, because there is no alternative. The daughter, María, who is betrothed to marry Ignacio, the plantation foreman, has plans of her own, however. Her aspiration, to migrate to the United States, seems to be within her grasp. Pepe, her tragic love interest, tells her that "behind the volcano is the United States. Well, there is Mexico in between." Her migration fantasy undermines plans for the wedding and potentially the future well-being of her entire family. What is significant to the discussion of justice advanced in this volume is first that the lure of the United States looms large, both as a paradise that would extricate the characters from their own difficult existences and a purveyor of all things of quality. For example, a pesticide is acquired from the United States to rid the fields of snakes, and Ignacio assures that this product will work because "it comes from the United States." To which his *compañero* replies, "So it must be good." The emphasis on the transnational context for even the most marginalized, rural Guatemalans reveals an important truth about the permeability of borders, even if only in the imagination. And second, most Guatemalans have very little interaction with the institutions of the state. In the film, María's family comes into contact with the state only once. It is a devastating encounter. María, who is in her final stages of pregnancy with Pepe's baby, after offering him sex in the hope that he would invite her to go with him to the United States, is taken to the hospital in the back of Ignacio's pick-up truck. The hospital staff cannot communicate with the family in their native language, Kaqchiquel, and no one in the family speaks Spanish. This leaves Ignacio, who is furious that his fiancée is pregnant by another man, to translate. In the truncated communications, Ignacio arranges for María's baby to be put up for adoption while she is unconscious, recovering from a nearly fatal snake bite. The adoption requires María's consent, but the family relies on Ignacio's explanation of the paperwork that is presented by hospital staff, and María signs the forms under the false pretense that the baby has died and they are requesting a government-subsidized coffin. The inability of the family to understand Spanish, the official language of the state, and the lack of translation services produce a layer of cultural violence that makes any measure of progress in the reform of state institutions virtually irrelevant. In addition, the cultural dimensions of gender and racialized oppression undermine the slowly strengthening structures of the state, which are connected to the war or are the enduring products of culture. They also function as parallel experiences of injustice. As Rita Palacios explains in chapter 7,

These formal processes categorize and archive the individual and collective memories that can be *recorded* but overlook what can only be *experienced*. In other words, what cannot be wholly addressed in formal legal proceedings, such as the inexpressible aspects of pain, fear, and loss, can be lost: the memories and feelings of the witness-survivor, for better or worse, become part of a larger narrative that defines truth and measures violence, relegating the unsayable to a distant memory, an uncomfortable silence, or a willed amnesia. And so, while reports, trials, and sentencing may bring justice, they are not enough to address the lasting effects of violence. (Palacios, this volume, 139)

These realities are the focus of Guatemala's post-peace generation of artists and cultural critics.

Yet the degree to which a country's cultures can be separated from its institutions is questionable. The main problem for Guatemala, as documented in this book, is that the institutions of the state produce consistently unjust outcomes in spite of monumental efforts to resist them. This failure exists in contradiction to both the significance of the symbolic effects of institutional successes, however modest, and the strengthening effects of testing the institutions of justice. These contradictions are replicated in the twentieth anniversary of the Peace Accords, which was met with both celebration and protest, the former boasting the achievements in post-war Guatemala and the latter insisting, "Nothing to Celebrate. Do not forget, nor forgive, nor reconcile" (Rodriguez 2016). This marks the ambivalence of justice in a culturally divided society that must rely on weak institutions for the promise of reconciliation. Much of this hope comes in the form of symbols – Peace Accords, court cases against Mynor Padilla, Hudbay, Tahoe Resources, Rios Montt, new institutions (CICIG), art, literature, and popular culture – that simultaneously stand for the promise of justice and the ways in which it is continually undermined. The meaning of these symbols is in flux, interpreted and reinterpreted by new generations of politicians, state officials, activists, scholars, and artists. The battle to assign new meanings to such symbols, or symbolic cases, becomes a struggle to alter the composition of cultural hegemony in Guatemalan society, where this term is used in its Gramscian sense of "the consensual basis of an existing political system within civil society" (Adamson 1980, 170). Weak states, Antonio Gramsci maintained, resort to violence to govern; strong states govern by hegemonic consensus alone. Guatemala, as noted above, is a weak state; yet the twenty-first-century international context does not allow it to resort to violence on the scale that it did from 1961 to 1996. This opens the terms of hegemonic consensus to competition.

Though the Ladino elite, bolstered by its private financial resources and its ties to foreign governments and corporations, remains dominant, "the symbiosis between foreign interests and the Guatemalan elite" (Wright 1989, 129) of the 1980s is no longer axiomatic. As countries in the Global North incorporate aspects of multiculturalism into their societies, they become less sympathetic to the assertion that Guatemala can modernize only if it is organized on a model of Ladino supremacy and "Western" values. Though post-1990 accelerated globalization threatens Mayan cultures with assimilation, it also provides them with tools, as will be argued below, to disseminate their cultural cosmovision, and even their languages, through the internet and networks of international supporters.

One source of hope in responding to persistent cultural violence is that some new artists from marginalized backgrounds, who work in accessible popular forms, have begun to break social taboos ingrained in Guatemala's historic power structures. These structures divide the country along the axes of cultural identification (Ladino-Indigenous) and language (Spanish-Mayan languages), which are mimetic of economic rifts and the urban-rural split. Altering the contours of cultural hegemony in ways that will promote the reform of Guatemala's massively skewed land distribution, which has persisted since the Spanish Conquest in 1524, and is the underlying structural source of the country's economic injustice,[1] requires change in the culture of the capital, where more and more Indigenous people live, but where the urban landscape does not acknowledge their contribution. Guatemala City, developed after the 1773 earthquake that destroyed the former capital of Antigua, is a city replete with symbols of Ladino domination. Like the most disenchanted reading of the "Las guerras son inútiles" slogan adopted to celebrate the twentieth anniversary of the Peace Accords, much of the architecture of the modern city that lies to the south of the colonial centre is a chorus of triumphalist symbols of the ideology of the late nineteenth-century liberalism that provided the intellectual justification for contemporary racism. Latin American liberalism, as exemplified by presidents such as Benito Juárez (1858–72) of Mexico and Domingo Faustino Sarmiento (1868–74) of Argentina, is remembered for its promotion of individualism, anti-clericalism, freedom of the press and free markets, in opposition to a conservative culture that promoted the restoration of eighteenth-century social hierarchies, the dominance of the Catholic Church in both social mores and the economic order, and a mercantilist economy that guided and regulated commercial activity from the summits of power. In every country where liberals took control, Indigenous peoples saw their rights to their ancestral lands legislated away to create a free market in real estate. This plunder was justified by the ideology of positivism, adopted from French thinkers' attempts to develop a scientific approach to the problems of society, but which, in the hands of ardent nineteenth-century Latin American modernizers, "absorbed the ideas of Social Darwinism, which posited a racial hierarchy in which whites were deemed superior to other races. In Latin America such doctrines relegated the majority of the population to an inferior status and were partly responsible for the attempts to encourage European immigration in many countries so as to whiten the population and improve the chances of progress" (Williamson 1992, 283–4). The Guatemalan liberal strongman Justo Rufino Barrios, who was president from 1873 to 1885, like other leaders influenced by positivism, regarded progress as inseparable from the suppression, assimilation, or annihilation of Indigenous cultures. Positivist beliefs paved the way for Guatemalan experiments in racial tinkering, notably the importation of Germans to the Alta Verapaz, Baja Verapaz, and Quetzaltenango regions in the years between Barrios's presidency and the early 1940s. Positivist ideology dictated that the Germans, as a "superior" people, would "improve the race" by intermarrying with the local Maya. In the Alta Verapaz region, in particular, Germans became the dominant group in society. Rather than intermarrying, however, they expelled the local Q'eqchi'-Maya from their traditional lands and established lucrative coffee plantations. The Germans "borrowed from German banks, owned their own transport facilities and fully integrated marketing and distribution networks. Few gave up their German citizenship or integrated into Guatemalan life" (Handy 1984, 66).

An equestrian statue of President Barrios dominates Plaza Barrios in the southern part of Guatemala City's historical centre of Zone 1; it is a major TransMetro transfer point that thousands of the capital's residents pass every day. Avenida La Reforma, Guatemala's widest street, was built during the period of liberal dominance. In the words of Rodrigo Rey Rosa, it is named after "la despiada reforma que abolió el derecho de los indígenas guatemaltecos a sus tierras comunales para que fueran convertidas en plantaciones de café ... avenida abierta, aplanada y pavimentada por los mismos indígenas cuyas tierras habían sido usurpadas por aquella reforma" (Rey Rosa 2001, 27; the heartless reform that abolished Guatemalan Indigenous peoples' rights to their communal lands so that they could be converted into coffee plantations ... an open avenue, paved and levelled by the same Indigenous people whose lands that reform had usurped). The result of Barrios's reforms was that by 1876 one in four Mayan men were working as forced labourers on a coffee plantation, often foreign owned (Stewart 2015, 318). Avenida La Reforma is the boundary between Zones 9 and 10, two of Guatemala City's most exclusive neighbourhoods; inevitably, and ominously, the U.S. Embassy looks out over this avenue. By the same token, Séptima Avenida in Zone 9 is straddled by the Torre del Reformador (Reformer's tower), a steel imitation of Paris's Eiffel Tower constructed in honour of Barrios and his political legacy. To celebrate the liberals' seizure of power on 30 June 1871, and the consolidation of Ladino supremacy produced by the reforms they implemented, a bell at the top of the tower is rung every year on 30 June.

As this book demonstrates, challenges to the myths of Ladino supremacy and "useless" wars issue from various sectors: Mayan elders, organized labour, NGOs, certain lawyers and judges, anti-mining activists, novelists, poets, performance artists. In recent years, popular music has played a growing role in broadcasting dissident interpretations of Guatemalan history, particularly in songs that become available as YouTube videos. The Ladina singer Rebeca Lane's snarling rap tune "Cumbia de la memoria" (2015; Cumbia of memory) insists "Sí, hubo genocidio" (Yes, there was a genocide). Lane draws explicit links between the 1980s and the abuses wreaked by mining in subsequent years, rhyming "matar mayoría" (killing the majority) with "en estas tierras ya hay minería" (there's now mining in these lands) (Lane 2015). In a more tongue-in-cheek vein, Fernando Scheel and Raquel Pajoc's Kaqchiquel-language "Pa' capital" (2014; To the capital) slyly insists on the growing Mayan presence in Guatemala City. "Pa' capital" became an online phenomenon, accumulating more than 40,000 views in the first week it was posted on YouTube. As the news and public affairs website Soy502.com noted, the song prompted debate "sobre varios temas que la sociedad necesita plantearse, desde la interculturalidad hasta la centralización política y económica del área metropolitana" (2014, "Raquel Pajoc"; on various subjects that society needs to consider, from interculturality to the political and economic centralization of the metropolitan area). The video, which shows young Mayan women in traditional dress walking and dancing in Guatemala City's Parque Central (Central Park) as they describe the capital in Kaqchiquel, inverts the centuries'-old hierarchy of Spanish being used to describe, often disparagingly, the rural areas inhabited by the majority of the Mayan population.

A dramatic inversion of traditional power structures occurred on 27 April 2016. For the first time in his distinguished career, the Spanish flamenco master Diego Jiménez, known as "El Cigala" (The prawn), gave a concert in Guatemala. The Miguel Ángel Asturias National Theatre was sold out. The Ladino elite, eager to assert their kinship with the culture of Spain, snapped up the high-priced tickets and arrived at the event garbed in their most expensive finery – only to discover that the opening act was a young Mayan folksinger dressed in a traditional *huipil* and *üq* (Mayan skirt), who sang in Kaqchiquel as well as in Spanish.[2] This may have been the first time since 1524 that the Ladino elite not only allowed a Mayan woman to address them from a stage, in her own language, but also applauded her.

The career of this young singer, Sara Curruchich Cúmez, who has become Guatemala's best-known new popular musician since the emergence of the Grammy-nominated (and apolitical) Ladina singer Gaby Moreno a decade earlier, illustrates the opportunities and limitations for advancing the debate about justice in the twenty-first century. Curruchich was born the fifth and youngest child of a Kaqchiquel-speaking family in San Juan Comalapa, Chimaltenango, in 1993. Comalapa, a town of 35,000 people located on a remote hilltop eighteen kilometres off the Panamerican Highway, suffered massacres during the civil war (Catherine Nolin, in chapter 2, describes the exhumation of bodies of the disappeared from the Comalapa military base). A white stone tablet at the entrance to the central downtown street lists the names of the town's disappeared, including members of both the Curruchich and Cúmez families, describing them as "víctimas del genocidio y de la represión" (victims of genocide and repression). This monument's inclusion of the word *genocide* makes it very unusual in Guatemala; in addition to this exceptional institutional recognition, a second peculiarity of Comalapa's history was instrumental in Sara Curruchich's formation. Styling itself "La Florencia de las Américas" (The Florence of the Americas), Comalapa is famous for its painters, who produce landscapes and scenes of rural Guatemalan life in a naive, picturesque style. The Comalapa school of painting was initiated around 1930 by Andrés Curruchich (1891–1969), a first cousin of the singer's grandfather, and was vaulted to cultural prominence under the government of Arbenz (Wright 1989, 152). The Comalapa school's most salient creations are the murals that line the walls along the approach road into town.

Reminiscent of Diego Rivera's murals in the National Palace of Mexico, yet deliberately simpler and blunter in execution, the Comalapa murals recount the history of the Kaqchiquel people. They include idealized images of the post-classical Kaqchiquel capital of Iximché, ugly depictions of Spanish conquistadors and Catholic priests, and gruesome scenes from the massacres of the 1980s. This combination of an acute historical consciousness of violent oppression and an environment where, in spite of pervasive poverty, the dream of living as an artist was imaginable, shaped Sara Curruchich's awareness and aspirations.

Curruchich's father, in addition to being an agricultural labourer and a carpenter, was an accomplished musician. Like the Kaqchiquel family in *Ixcanul*, the Curruchiches were forced by economic necessity to leave their home region to work in other parts of Guatemala. As a child, Sara sold oranges to cars stuck in traffic on Guatemala City's Avenida Roosevelt, an experience that introduced her to "la discriminación y la exclusión" (Escobar Sarti 2016; discrimination and exclusion). Her father began to teach her the guitar when she was five; she played and sang with him daily until her adolescence. She got her break in March 2014 when the Dresden Philharmonic Orchestra asked her to sing with them in Mexico City.[3] The German musicians were so impressed by Curruchich's deep, resonant voice that they offered to record, and make a video of, one of her songs. The result, "Ch'uti'xtän" (Curruchich Cúmez 2015a; Little girl), accumulated more than 400,000 views on YouTube and became a staple of Guatemalan radio airplay. The popularization of a song whose chorus was in Kaqchiquel, albeit framed by an introduction and conclusion in Spanish, represented a cultural watershed. The structure of Currichich's songs adumbrates the rebalancing of cultural hegemony that might characterize a more just social order. Spanish and Mayan words cohabit; Curruchich neither retreats into purely Mayan lyrics that her community alone will understand, nor becomes a Spanish-language singer: she insists on the participation of Kaqchiquel words in national debates, on Mayan cultures as contemporary and not folkloric, and on the intercultural essence of Guatemalan identity. Though one of her best-known songs, "Ixoqi" (Curruchich Cúmez 2015b; Women), is entirely in Kaqchiquel, and a number of others, particularly those on activist themes, are in Spanish only, her preferred mode appears to be to mix the languages. This insistence that Guatemala can be understood only by pairing Spanish and Mayan perspectives is not overtly politicized in "Ch'uti'xtän," which tells the story of how her parents fell in love. Yet it is noticeable that in this narration of two children meeting "en una fiesta tradicional" (Curruchich Cúmez 2015a; at a traditional celebration), and growing up to marry, Spanish yields to Kaqchiquel at the point where strong emotion intervenes. The key lines, which are both intimate and reflective of cultural tradition – "Katk'ule' wik'in wi rat yinawajo'" (If you love me, marry me) – are in Kaqchiquel.

The quest for justice for Mayan peoples is central to Curruchich's music. "Ch'uti'xtän" is an anomaly in her repertoire because, in contrast to most singers with high public profiles, she rarely sings about love. The first song she composed when she resumed her musical career at eighteen, "Amigo," is dedicated to an activist friend who was murdered. The men in her songs are not identified as lovers, but rather as what politically engaged Central Americans of the 1980s called "compañeros de lucha" (comrades in struggle). Her songs emphasize collective destiny rather than personal fulfilment. This thematic tendency has strengthened Curruchich's fan-base in the activist, student, and Indigenous communities, both nationally and internationally, yet has held back her conquest of a mass audience in Guatemala. Following on the success of "Ch'uti'xtän," Curruchich's single "Resistir" was released in March 2016 with the support of a substantial media campaign and was used to launch her international tour. The song is entirely in Spanish, yet it received no radio airplay because the lyrics were deemed too political, even though they contain no reference to any specific political event. The lyrics urge listeners to continue struggling, regardless of the hardships they encounter. The song opens with the words, "Que no se mueran los sueños, que no se apague la luz," (May dreams not die, may the light not be extinguished) and moves, via reminders that our lives leave tracks behind them, and references to an ultimate "victoria" (victory), to the refrain, "Que no se apague nuestro amor" (Curruchich Cúmez 2016; May our love not be extinguished). As in "Amigo," the love expressed here is that of a collectivity. Though the song is expertly crafted, the language of the lyrics echoes 1980s Central American revolutionary discourse – the promise of an eventual victory, the notion of a love for the people that surpasses the egotism of love for an individual – in ways that unnerved mainstream radio programmers. Curruchich, who wrote "Resistir" to boost the morale of Mayan communities that are defending their land against mining companies and hydroelectric projects, is unyielding in her conviction that the achievement of social justice requires radio programming to move beyond the paradigm of romantic love: "Quieren seguir vendiendo [el amor romantico] y nos quieren tener sometidos en esto, pero no quieren que tengamos algún tipo de conciencia y de valorarnos y de encontrarnos a nosotras y nosotros mismos" (Curruchich Cúmez 2017a; They want to keep selling us romantic love and they want to keep us submissive through that, but they don't want us to have any kind of consciousness or to value each other or to get to know each other, women and men both). The collectivity that Curruchich evokes in opposition to the egotistical search for fulfilment that distracts individuals from their fraying communities is that of Mayan traditionalism in which neighbours said, "seguro que van a tapiscar la cosecha la próxima semana. Yo les ayudo, y cuando tenga que hacer mi cosecha, ustedes vienen, o si yo tengo que desgranar el maíz…, ustedes me ayudan" (Curruchich Cúmez 2017a; I see you're going to harvest your crops next week. I'll help you, and when I have to do my harvest, you'll come, or if I have to thresh my corn…, you'll help me). Her complaint about contemporary Guatemalan society is not only that it is unjust, but also that it is excessively materialistic and tells the individual, "No vas a ser feliz si no tienes un carro o dos o tres teléfonos" (You're not going to be happy if you don't have a car or two or three telephones).

"Resistir" launched Curruchich's 2016 international tour, an unprecedented event for a Mayan musician, which included an appearance at the United Nations in New York, and concerts in many areas of Guatemala, as well as in Madrid, Barcelona, Berlin, Paris, Brooklyn, Biarritz, and Austin, Texas. This elevation of Curruchich's vision of Guatemala's history and struggle for justice to a global level illustrates how, in the twenty-first-century, Indigenous artists can make an end-run around the Ladino power structures that dominate the nation. The Dresden Philharmonic Orchestra launched Curruchich's career, the French film director Jean-Stéphane Sauvaire secured her invitation to the United Nations, and her French manager, Vincent Simon, organized her tour. These foreign actors cannot alter Guatemala's power structures, but they are able to exert pressure from without by diffusing Curruchich's politicized, Mayan-centred vision of Guatemala to an international audience. By contrast, when Ladinos situate Curruchich, they render her folkloric and politically anodyne. "No se olvida de donde se viene" (2015c; One doesn't forget where one comes from), the video that celebrates Curruchich's 2015 Guatemalan Artist of the Year Award, sponsored by Banco Industrial, looks like a tourism clip, with extended shots of the ruins of Tikal, children playing marimbas, and light-skinned Ladinos waving Guatemalan flags in a show of national pride.

The lyrics evoke most of the clichés of Guatemala as a friendly, beautiful land of eternal spring while avoiding any reference to the political consequences of where one comes from. Curruchich is allowed one line each in Kaqchiquel at the song's beginning and end; her closing line is followed by a male voiceover that reinterprets her words by urging viewers to "construir la Guatemala con la que soñamos" (Curruchich Cúmez 2016c; build the Guatemala we dream of) as the final panning shot of office blocks fronted by a brightly lighted freeway illustrates the kind of Guatemala of which Banco Industrial dreams.

The national conundrum, in which powerful Ladinos negate Curruchich's quest to articulate historical awareness, can be obviated through alliances with foreigners or Ladino progressives.[4] The far greater ease of collaboration with such figures in the twenty-first century provides Curruchich with an advantage that activists of earlier decades did not enjoy. The documentary filmmaker Pamela Yates, director of works on Guatemalan themes such as *When the Mountains Tremble* (1983) and *Granito: How to Nail a Dictator* (2011), which was presented as evidence at the Ríos Montt trial, attended Curruchich's Brooklyn concert. She heard the then-unrecorded "Ralk'wal Ulew" (Daughters and Sons of the Earth)[5] and asked to use it as the musical introduction to *500 Years* (2017), the third film in her Guatemalan trilogy. The lush production values of the video supplement the song's words, which draw on Mayan religious concepts to provide an emotionally intense elegy to the victims of genocide. The narrator states that the dead have "transcendidos" (Curruchich Cúmez 2017b; transcended) this life and are now in the heart of the sky and the heart of the earth; she promises that, wherever the dead may be, she will ask the wind to whisper in their ear. The video rehearses the passage from the mourning of rural communities in the aftermath of genocide to the Ríos Montt trial to the 2015 demonstrations that forced Ríos Montt's former deputy during the genocide, Otto Pérez Molina, to resign as president. The video's sequencing deliberately creates the illusion of the crowds spilling out of Ríos Montt's trial and acting on their renewed historical consciousness by pouring down Sexta Avenida, the pedestrianized avenue at the core of Mayor Álvaro Arzú's recreation of the city's historical centre as an attractive modern destination devoid of echoes of past violence, to the Parque Central to demand Pérez Molina's resignation (as Candace Johnson points out in chapter 1, the president's resignation was due to charges of Customs fraud, not genocide). A brief shot of the 1980s Mayan human rights icon Rigoberta Menchú Tum, winner of the 1992

Nobel Peace Prize, now in late middle age, smiling up at the protesters, underscores the debt the struggles of the present owe to an awareness of the past. The video portrays a consciousness of history nullifying cosmetic attempts to change the landscape where genocide occurred, while the lyrics of "Ralk'wal Ulew," entirely in Kaqchiquel for the first half of the song, vow that the victims of genocide will be not forgotten.

Curruchich sings in defiance of the cultural violence that contributors to this volume view as more recalcitrant than the structural variety. One central goal of her music is to keep alive debates surrounding injustice. While some speak of a Guatemalan "despertar" (awakening) in 2015, she emphasizes that "tenemos que seguir despertando, no despertar y dormirnos otra vez" (Curruchich Cúmez 2017a; we have to keep on awakening, not wake up and go to sleep again). Having been warned by her mother not to participate in the marches against Pérez Molina (advice she ignored), she is aware of the fear that continues to paralyse the older generations traumatized by their experiences of disappearances and massacres. Her resuscitation of Kaqchiquel as a language in which national debates may be sung about is complex. In New York, Paris, and Berlin, singing in Kaqchiquel reaffirms foreign views of Guatemala as the land of the Maya; inside Guatemala, using in public a language that has been the private currency of uneducated rural people, particularly women who work outside the cash nexus where Spanish dominates, and mixing this language with Spanish to assert that the two are of equal legitimacy, is a challenge to racism. Speaking of the shame that makes some Indigenous parents reluctant to teach Mayan languages to their children, she says, "No se puede hablar de sentir vergüenza sin enlazarlo con la discriminación, y enlazarlo con el racismo de institución que se ha tenido hacia las personas que son mayahablantes" (a; You can't speak about feeling ashamed without connecting it to discrimination, and connecting it to the institutional racism that there has been towards people who are Mayan-speaking). For the Maya, more than anyone else, a generalized understanding of the abuses and institutional violence that continue to form part of the engine of injustice in Guatemala is dependent on a recognition of the genocidal acts of the 1980s: "No vamos a tener justicia y no vamos a tener paz y no vamos a tener esa memoria de todo eso, y entonces no podemos hablar de cultura, no podemos hablar de historia, cuando … muchos grupos no reconocen que sí hubo un intento de arrasar con toda la población indígena" (a; We're not going to have justice and we're not going to have peace and we're not going to have that memory of all that, and then we can't talk about culture, we can't talk about history when … many groups do not recognize that yes, there was an attempt to wipe out the whole Indigenous population).

In one of her first interviews, given when she was still singing with the group Sobrevivencia, Curruchich observed that, though her community supported her career, some considered it wrong for a Mayan woman, traditionally the most reticent figure in Guatemalan society, to be standing on stage expressing her feelings through song. Some Maya made comments such as: "Usted no debería estar acá" (Curruchich Cúmez 2014; You shouldn't be here). For both Maya and Ladinos, Curruchich's career represents a cultural upheaval. Even a Mayan woman as committed to public interventions as Rigoberta Menchú maintained that the Mayas' strength came from keeping their secrets (Burgos-Debray 1982, 271). Employing an international audience, and the internet, as a single young woman, to defend a rural, agricultural, animist, patriarchal culture brings Curruchich into contradiction with her own heritage, notably in its gender roles; the more visible she becomes in the cause of defending her culture, the more she changes it. This kind of exposure, which contravenes Mayan tradition, brings the "contemporary" and "folkloric" versions of Mayan culture into conflict and sometimes causes her to feel shame. Referring to the unexpected success of "Ch'uti'xtän," she recalls: "Yo una vez iba de la Ciudad a Comalapa, y en la camioneta sonó, y me sentí – fue tan extraño que me dio mucha alegría, me dio no tristeza sino como vergüenza, en la camioneta, qué pena, ¿no?" (Curruchich Cúmez 2017a; Once I was on my way from Guatemala City to Comalapa and in the bus ["Ch'uti'xtän"] came on and I felt – it was so strange that it made me very happy, it made me feel not sadness but rather shame, right there in the bus, what a pity, no?)

The example set by Sara Curruchich will liberate the next generation of Mayan women to play a more public role, enlarging the constituency that participates in future debates about justice. Technology, too, has enlarged this constituency. On her 2016 tour, Curruchich was startled to find that after each concert, regardless of the country she was in, she was approached by Guatemalans working abroad, many of them living in conditions of financial or legal precarity: "no era frustrante, sino agobiante" (Curruchich Cúmez 2017a; it wasn't frustrating so much as overwhelming). Yet, thanks to WhatsApp and Skype, these Guatemalans abroad, whose remittances support 38 per cent of the country's population (Gamarro 2017), are in increasingly intimate contact with the nation they left behind. They bring their experiences in other societies

to bear on Guatemalan debates. In the collapsible time frame of Guatemalan history, where progress sometimes stagnates and sometimes
leaps forward, the persistence of Ladino racism, widespread malnutrition, unequal access to clean water, proposals to construct hydroelectric dams in Mayan communities, constitutional reform (including the
proposed formal recognition of Mayan systems of justice), reproductive
health issues, and the enduring dilemma of skewed land distribution
will be confronted by a Guatemalan public that increasingly includes
those who live overseas as well as those who reside in Guatemala. The
nation, like everything else, has been globalized, yet the conditions of
national life affect those who live there. When the thirtieth anniversary of the Peace Accords is celebrated, the government of the day will
need to devise a slogan that is more meaningful and respectful of the
Guatemalan people's experience than the empty assertion that wars are
useless.

NOTES

1 After the signing of the Peace Accords, many Mayan leaders assumed that
  the country's unequal land distribution would dwindle in importance
  as the economy modernized. In 2003, Rigoberto Quemé Chay, the Maya-
  Quiché mayor of Quetzaltenango and a potential presidential candidate,
  stated, "The children of Mayan peasants don't want to sow corn ... As we
  become more connected to global markets, productivity depends more on
  artisans, on a productive fringe of small- and medium-sized businesses. In
  time the land issue is going to become less important" (Henighan 2003). The
  advent of mining companies, and the failure of the Guatemalan economy
  to break large swathes of the population's dependence on near-subsistence
  agriculture, however, have reinstated land tenure as a pivotal political issue.
2 The Ladino assumption that flamenco symbolizes Europeanness and white-
  ness is, of course, erroneous. In Spain, most flamenco artists come from the
  south; many, like Diego Jiménez, are brown-skinned men of Roma ancestry.
3 When her father died of a neurological disorder, Sara Curruchich fell into
  a prolonged depression and abandoned music. After five or six years of
  silence, she began to play and sing again at the age of eighteen, determined
  to make music her career. Between 2011 and 2014, Curruchich qualified as
  primary-school music teacher, started a degree in music at Universidad de
  San Carlos de Guatemala (USAC: San Carlos University) but dropped out
  for lack of funds, joined the Mam-Mayan rock band Sobrevivencia, and

taught music in primary schools for a year each in Comalapa and Guatemala City.

4 In March 2017 Curruchich produced a video with the inner-city rapper Kontra (whose legal name is Danny Marín), which the two singers presented as an attempt to overcome the alienation that divides Ladinos from Mayas (Curruchich Cúmez and Kontra 2017). In the same month she presented a concert with the feminist, anarchist Ladina rapper Rebeca Lane.

5 The Kaqchiquel "ak'wal" (or "alk'wal") is usually translated as "children." A literal rendering of the Kaqchiquel would be "children of the earth." Curruchich translates the song's title into Spanish as "Hijas e hijos de la tierra" (Daughters and sons of the earth), emphasizing Mayan genocide survivors' living relationship with the planet, adding a gender differentiation that is not present in the Kaqchiquel and, as she does in both her lyrics and in conversation, refusing to use the Spanish masculine plural as a generalized form, but rather including both the feminine and masculine forms, and placing the feminine first.

WORKS CONSULTED

Adamson, Walter L. 1980. *Hegemony and Revolution: A Study of Antonio Gramsci's Political and Cultural Theory*. Berkeley: University of California Press.

Associated Press. 2017. "Guatemala Arrests Supreme Court Justice on Corruption Charge." 8 February. https://www.yahoo.com/news/guatemala-arrests-supreme-court-justice-corruption-charge-202248799.html.

Bernard, Renee. 2017. "BC Court of Appeal Says Guatemalan Protesters Can Have Lawsuit Heard in BC." News 1130, 26 January. http://www.news1130.com/2017/01/26/bc-court-appeal-says-guatemalan-protesters-can-lawsuit-heard-bc/.

Burgos-Debray, Elisabeth, ed. 1982. *Me llamo Rigoberta Menchú y así me nació la conciencia*. Guatemala: Arcoiris.

Bustamante, Jayro, dir. 2015. *Ixcanul*. With María Telón, Justo Lorenzo and Marvin Coroy. La Casa de Producción / Tu Vas Voir.

Curruchich Cúmez, Sara. 2014. "Entrevista." 31 July. https://www.youtube.com/watch?v=wz6-nALNNsE.

– 2015a. "Ch'uti'ixtän." 17 January. https://www.youtube.com/watch?v=n0E1efv6mLA.

– 2015b. "Ixoqi." 1 January. https://www.youtube.com/watch?v=FRETLDOA-Pw.

– 2015c. "No se olvida de donde se viene." 28 September. https://www.youtube.com/watch?v=jYUMm8BRjY8.
– 2016. "Resistir." 1 March. https://www.youtube.com/watch?v=5lP9duv_7xc.
– 2017a. Interview with Stephen Henighan, Casa de Cervantes, Guatemala City, 22 February.
– 2017b. "Ralk'wal Ulew." 18 January. https://www.youtube.com/watch?v=D5iAusanYkg.
Curruchich Cúmez, Sara, and Kontra. 2017. "Ser del viento." 7 March. https://www.youtube.com/watch?v=ulIhmsLVRsw.
Escobar Sarti, Carolina. 2016. "Sara Curruchich, mujer de voz con alas." http://www.saracurruchich.com/Templates/bio.dwt.php.
Gamarro, Urías. 2017. "Remesas benefician al 38% de guatemaltecos." *Prensa Libre*, 17 February. https://www.pressreader.com/guatemala/prensa-libre/20170217/281522225847130.
Goldman, Francisco. 2007. *The Art of Political Murder: Who Killed the Bishop?* New York: Grove.
Handy, Jim. 1984. *Gift of the Devil: A History of Guatemala*. Toronto: Between the Lines.
Henighan, Stephen. 2003. "Majority No Longer Silent: Maya Increase Political Involvement." *Gazette (Montreal)*, 17 June.
Lane, Rebeca. 2015. "La Cumbia de la Memoria." 13 June. https://www.youtube.com/watch?v=7bw_0e__U6k.
Martín, Sabrina. 2017. "Guatemala President Halts Promise to Donate 60 Percent of His Salary amid Legal Issues." *PanAm Post*, 30 January. https://panampost.com/sabrina-martin/2017/01/30/guatemala-president-halts-promise-to-donate-60-percent-of-his-salary-amid-legal-issues/.
Morales, Jimmy. 2016. "Discurso del Presidente de la República, Señor Jimmy Morales Cabrera con Motivo de la Conmemoración de los XX años de la Firma del Acuerdo de Paz Firme y Duradera." COPREDEH. http://copredeh.gob.gt/discurso-del-presidente-la-republica-senor-jimmy-morales-cabrera-motivo-la-conmemoracion-los-xx-anos-la-firma-del-acuerdo-paz-firme-duradera/.
Mullen, Matthew. 2015. "Reassessing the Focus of Transitional Justice: The Need to Move Structural and Cultural Violence to the Centre." *Cambridge Review of International Affairs* 28 (3): 462–79.
Oettler, Anika. 2006. "Encounters with History: Dealing with the 'Present Past' in Guatemala." *Revista Europea de Estudios Latinoamericanos y del Caribe* 81 (81): 3–19. https://doi.org/10.18352/erlacs.9645.

"Raquel Pajoc, la protagonista del video 'Pa' Capital.'" 2014. *Soy502*, 7 October. www.soy502.com/articulo/raquel-pajoc-protagonista-video-pa-capital.

Rey Rosa, Rodrigo. 2012. *Los sordos*. D.F., México: Alfaguara.

Rodriguez, James. 2016. "Guatemala Commemorates 20th Anniversary of Peace Accords." Telesur, 31 December. https://www.telesurtv.net/english/news/Guatemala-Commemorates-20th-Anniversary-of-Peace-Accords-20161231-0009.html.

Scheel, Fernando, and Raquel Pajoc. 2014. "Pa' capital." 1 October. https://www.youtube.com/watch?v=N4evy_IMFKQ.

Stewart, Iain, ed. 2015. *The Rough Guide to Guatemala*. London: Rough Guides.

Taylor, Susan. 2017. "B.C. Court Rules Guatemalan Lawsuit against Tahoe Resources Can Proceed." *Globe and Mail*, 26 January. https://www.theglobeandmail.com/news/british-columbia/bc-court-rules-guatemalan-lawsuit-against-tahoe-resources-can-proceed/article33790534/.

Wall Street Journal. 2017, "Guatemalan President's Brother, Son Detained in Corruption Probe." 19 January. https://www.wsj.com/articles/guatemalan-presidents-brother-son-detained-in-corruption-probe-1484796905.

Williamson, Edwin. 1992. *The Penguin History of Latin America*. London: Penguin.

Wright, Ronald. 1989. *Time Among the Maya: Travels in Belize, Guatemala and Mexico*. Markham, ON: Penguin Canada.

Contributors

Kalowatie Deonandan is professor of political studies at the University of Saskatchewan. She has co-edited three volumes: *Undoing Democracy: The Politics of Electoral Caudillismo in Nicaragua* (Lexington Books, 2004), *Revolutionary Movements to Political Parties: Case Studies from Latin America and Africa* (Palgrave, 2007), and *Mining in Latin America: Critical Reflections on the New Extraction* (Routledge 2016). She has over fifteen years' experience teaching and doing research in Guatemala. She is researching gender, mining in Latin America, and nuclear sector development in Canada.

Stephen Henighan is professor of Spanish and Hispanic Studies at the University of Guelph. A finalist for the Governor General's Literary Award and the Canada Prize in the Humanities, he has taught, done research, and worked as a freelance journalist in Guatemala. His publications on Central America include *Assuming the Light: The Parisian Literary Apprenticeship of Miguel Ángel Asturias* (Legenda, 1999), *Sandino's Nation: Ernesto Cardenal and Sergio Ramírez Writing Nicaragua, 1940–2012* (McGill-Queen's University Press, 2014), and *The Path of the Jaguar* (Thistledown, 2016), a novel set in Guatemala.

Candace Johnson is professor of political science at the University of Guelph. She is the author of *Health Care, Entitlement, and Citizenship* (University of Toronto Press, 2002) and *Maternal Transition: A North-South Politics of Pregnancy and Childbirth* (Routledge, 2014). She was the winner of the Jill Vickers Prize from the Canadian Political Science Association in 2009 and 2017, and has taught and done research in Guatemala.

W. George Lovell is professor of geography at Queen's University. He has made more than sixty research trips to Guatemala, and is author, co-author, or co-editor of more than a dozen books about the country. He is the long-time co-editor of the journal *Mesoamérica*. His research focuses on sixteenth- and seventeenth-century Guatemala. His study of the Spanish Conquest of the country, *Conquest and Survival in Colonial Guatemala* (McGill-Queen's University Press, 1992), is in its fourth revised edition. His book on modern Guatemala, *A Beauty That Hurts: Life and Death in Guatemala* (University of Texas Press, 1995), is in its second revised edition.

Helen Mack Chang was born in southwestern Guatemala and pursued a career as a businesswoman until the murder of her sister, the anthropologist Myrna Mack, in 1990. Since 1993, Helen Mack Chang has directed the Myrna Mack Foundation, which secured prosecution of her sister's murderers and also of their commanding officers, pioneering the concept of "intellectual authorship" of a crime at the Inter-American Court of Human Rights. The Myrna Mack Foundation works for human rights and victims' support programs in Guatemala, and has served in a human rights consulting role with many foreign governments. Helen Mack Chang has received various honorary doctorates and human rights awards, including Sweden's Right Livelihood Award.

Lisa Maldonado is a freelance translator of Spanish and French based in Guelph, Ontario. She lived in Guatemala for more than ten years.

Catherine Nolin is associate professor and chair of geography at the University of Northern British Columbia. She is the author of *Transnational Ruptures: Gender and Forced Migration* (Ashgate, 2006) and co-author with Jennifer Reade of *Empowering Women: Community Development in Rural Guatemala* (VDM Verlag, 2008). Nolin has won numerous teaching awards and has organized six field schools for undergraduate students in Guatemala.

Rita M. Palacios teaches Spanish at Conestoga College in Kitchener, Ontario. She was previously assistant professor of Spanish at California State University at Long Beach and Concordia University in Montreal. Born in Guatemala, she immigrated to Canada with her parents as a child. Her book, *Unwriting Literature: A Proposal for Reading Maya*

Literature through Ts'íib, co-written with Paul Worley, will be published by University of Arizona Press in 2019.

Claudia Paz y Paz was born in Guatemala and earned a doctorate in law from the University of Salamanca, Spain. From 2010 to 2014 she was the first woman to serve as attorney general of Guatemala. During her mandate, record numbers of drug traffickers were imprisoned, and Guatemalan military officers responsible for the massacre of Indigenous Mayan people were brought to trial. She has been a visiting researcher at Georgetown University. In 2015–16, she was a member of the five-member team appointed by the Inter-American Commission on Human Rights to investigate the disappearance of forty-three students in Ayotzinapa, Mexico. She is secretary for multidimensional security at the Organization of American States.

Magalí Rey Rosa, born in Guatemala City, has worked as an environmentalist since 1983. She has participated in the creation of two important Guatemalan nature reserves and has founded a number of influential organizations, including Savia: School of Ecological Thought. Her column on the environment appeared in the Guatemala City daily newspaper *Prensa Libre* from 1996 to 2015. Rey Rosa has spoken on environmental issues in Central America, particularly the impact of mining, at conferences around the world.

Rebecca Tatham is a PhD candidate in the Department of Political Studies at the University of Saskatchewan. She has lived and worked for a non-profit organization in Guatemala for several years. Her research focuses on gender and nuclear development in Saskatchewan and the gendered dynamics of anti-mining struggles in Guatemala.

Index

Lightning Source UK Ltd.
Milton Keynes UK
UKHW040606061218
333434UK00001B/39/P